Ionic Liquids in Polymer Systems

ACS SYMPOSIUM SERIES **913**

Ionic Liquids in Polymer Systems

Solvents, Additives, and Novel Applications

Christopher S. Brazel, Editor
The University of Alabama

Robin D. Rogers, Editor
The University of Alabama

Sponsored by the ACS Division
Polymer Chemistry, Inc.

American Chemical Society, Washington, DC

Library of Congress Cataloging-in-Publication Data

Ionic liquids in polymer systems : solvents, additives, and novel applications / Christopher S. Brazel, editor, Robin D. Rogers, editor.

    p. cm.—(ACS symposium series ; 913)

    Includes bibliographical references and index.

    ISBN 0–8412–3936–3 (alk. paper)

    1. Polymer solutions—Congresses. 2. Ionic solutions—Congresses. 3. Solvents—Congresses. 4. Molecular dynamics—Congresses

    I. Brazel, Christopher S., 1970- II. Rogers, Robin D. III. American Chemical Society. Division of Polymer Chemistry, Inc. IV. Series.

QD381.9.S65I55   2005
547′.70454—dc22   2005040959

The paper used in this publication meets the minimum requirements of American National Standard for Information Sciences—Permanence of Paper for Printed Library Materials, ANSI Z39.48–1984.

PRINTED IN THE UNITED STATES OF AMERICA

Foreword

The ACS Symposium Series was first published in 1974 to provide a mechanism for publishing symposia quickly in book form. The purpose of the series is to publish timely, comprehensive books developed from ACS sponsored symposia based on current scientific research. Occasionally, books are developed from symposia sponsored by other organizations when the topic is of keen interest to the chemistry audience.

Before agreeing to publish a book, the proposed table of contents is reviewed for appropriate and comprehensive coverage and for interest to the audience. Some papers may be excluded to better focus the book; others may be added to provide comprehensiveness. When appropriate, overview or introductory chapters are added. Drafts of chapters are peer-reviewed prior to final acceptance or rejection, and manuscripts are prepared in camera-ready format.

As a rule, only original research papers and original review papers are included in the volumes. Verbatim reproductions of previously published papers are not accepted.

ACS Books Department

Contents

Preface...ix

1. Free Radical Polymerization of Styrene and Methyl
   Methacrylate in Various Room Temperature Ionic Liquids................1
   Hongwei Zhang, Kunlun Hong, and Jimmy W. Mays

2. Free Radical Polymerization of Methacrylates in Ionic Liquids.........17
   Veronika Strehmel, André Laschewsky, Heide Kraudelt,
   Hendrik Wetzel, and Eckhard Görnitz

3. Polymer Synthesis in Ionic Liquids: Free Radical
   Polymerization in Water-Soluble Systems..37
   Carlos Guerrero-Sanchez, Frank Wiesbrock, and Ulrich S. Schubert

4. Reverse Atom Transfer Radical Polymerization of Methyl
   Methacrylate in Ionic Liquids...51
   Hongyang Ma and Xinhua Wan

5. Applying Ionic Liquids for Controlled Processing of Polymer
   Materials..71
   John D. Holbrey, Ji Chen, Megan B. Turner, Richard P. Swatloski,
   Scott K. Spear, and Robin D. Rogers

6. Novel Solid Polymer Electrolytes Composed of Zwitterionic
   Liquids and Polymerized Ionic Liquids..89
   Hiroyuki Ohno, Satoko Washiro, and Masahiro Yoshizawa

7. Ionic Liquids as Alternative Plasticizers for Poly(vinyl
   chloride): Flexibility and Stability in Thermal, Leaching,
   and UV Environments...103
   Mustafizur Rahman, Hugh W. Shoff, and Christopher S. Brazel

8. Ion Gels Prepared by In Situ Radical Polymerization of Vinyl
   Monomers in Room Temperature Ionic Liquids as Novel
   Highly Conductive Polymer Electrolytes...119
   Md. Abu Bin Hasan Susan, Akihiro Nada, and
   Masayoshi Watanabe

9. Cross-Linked Polymers in Ionic Liquids: Ionic Liquids
    as Porogens...133
    Peter Snedden, Andrew I. Cooper, Yaroslav Z. Khimyak,
    Keith Scott, and Neil Winterton

10. Synthesis, Characterization, and Application of Cross-
    Linked Poly(ethylene glycol) Networks Used for the Gelation
    of Ionic Liquids..149
    Marc A. Klingshirn, Scott K. Spear, John D. Holbrey,
    Jonathan G. Huddleston, and Robin D. Rogers

11. Novel Soft Composite Materials Composed of Ionic Liquids
    and Single-Walled Carbon Nanotubes...163
    Takanori Fukushima and Takuzo Aida

12. Application of Trialkylimidazolium Liquids and Salts to the
    Preparation of Polymer-Layered Silicate Nanocomposites
    and Polymer-Carbon Nanotube Nanocomposites........................175
    Douglas M. Fox, Severine Bellayer, Marius Murariu,
    Jeffrey W. Gilman, Paul H. Maupin, Hugh C. De Long,
    and Paul C. Trulove

Indexes

Author Index..191

Subject Index...193

Preface

Ionic liquid (IL) research has been brought to the forefront of technologies, particularly in connection with green chemistry and engineering. The new capabilities allowed by ILs extend into disciplines that cross the entire spectrum of chemistry, as they provide novel solvents to carry out reactions, new media for extractions and other separations, and active agents for the development of electrolytic materials. Because the breadth of IL research crosses many barriers, it is important that the knowledge gained from the IL-specific symposia (such as offered by the American Chemical Society (ACS) in 2001 and 2003) be shared with researchers in the specific disciplines where they compete with existing technologies. It is in this light that the *Ionic Liquids in Polymers* symposium was assembled during the 227th National Meeting of the ACS in Anaheim, California, in March–April 2004. They symposium was hosted by the ACS Division of Polymer Chemistry, Inc.

By bringing together polymer scientists who also work with ILs, the remarkable advances in this field can be communicated and challenged, with the results presented spurring new ideas and a deeper understanding of the subject. The body of work represented in this book represents a collection of research at a time where ILs are just beginning to be applied to polymeric systems. The works range from ILs in free radical polymerization and novel additives for materials with superior properties to ILs as polymer solvents and porogens to assist in designing unique molecular structures. Because ILs represent a wide-range of new chemicals, many researchers are looking at traditional materials problems from a new perspective, while others are taking ILs directly into more recent fields such as reversible addition fragmentation chain transfer (RAFT) polymerizations. The work represented in this book includes contributions by researchers across the world who are at the forefront of IL technologies.

Christopher S. Brazel
Department of Chemical and Biological Engineering
The University of Alabama
Tuscaloosa, AL 35487–0203
(205) 348–9738 (telephone)
(205) 348–7558 (fax)
CBrazel@coe.eng.ua.edu

Chapter 1

Free Radical Polymerization of Styrene and Methyl Methacrylate in Various Room Temperature Ionic Liquids

Hongwei Zhang[1], Kunlun Hong[2], and Jimmy W. Mays[1–3,*]

Departments of [1]Material Science and Engineering and [3]Chemistry, University of Tennessee, Knoxville, TN 37996
[2]Chemical Science Division, Oak Ridge National Laboratory, Oak Ridge, TN 37831

Conventional free radical polymerization of styrene and methyl methacrylate was carried out in various room temperature ionic liquids (RTILs). The RTILs used in this research encompass a wide range of cations and anions. Typical cations include imidazolium, phosphonium, pyridinium, and pyrrolidinium; typical anions include amide, borate, chloride, imide, phosphate, and phosphinate. Reactions are faster and polymers obtained usually have higher molecular weights when compared to polymerizations carried out in volatile organic solvents under the same conditions. This shows that rapid rates of polymerization and high molecular weights are general features of conventional radical polymerizations in RTILs. Attempts to correlate the polarities and viscosities of the RTILs with the polymerization behavior fail to yield discernible trends.

Introduction

The first room temperature ionic liquid (RTIL): [EtNH$_3$][NO$_3$] (mp: 12°C) was reported in 1914 (*1*) and since then a great deal of research effort has been exerted to explore the possible applications of RTILs. Initially, they were used mainly as electrolytes in batteries or for metal electrodepositions.

RTILs are also called molten salts and are liquids containing only cations and anions at ambient temperature. They exhibit a relatively wide electrochemical window, good electronic and ionic conductivity, a broad range of room temperature liquid compositions, negligible vapor pressure and excellent chemical, thermal, air and, in many cases, moisture stability (*2*). They are composed of weakly coordinating anions, *e. g.* BF$_4^-$ and PF$_6^-$ and, hence, are highly polar yet non-coordinating solvents. Most imidazolium and pyridinium ionic liquids have polarities similar to those of lower alcohols (*3*). Their hydrophilicity/lipophilicity are adjustable by varying the combination of cations and anions and, therefore, they are able to dissolve a variety of organic, inorganic and organometallic compounds. For this reason, RTILs have been referred to as "designer solvents" (*4*). Their ease of handling and potential for recycling make them attractive potentially environmentally benign to replace deleterious volatile molecular solvents.

Furthermore, RTILs have limited miscibility with some of the common organic solvents but high compatibility with transition metals. As a consequence, a biphasic or phase-separable (organic/RTIL biphasic systems) catalysis concept can be developed in which a homogeneous catalyst is immobilized in one liquid phase (RTIL) and the reactants and/or products reside largely in another liquid phase (organic) (*5*), thus enabling easy product and catalyst separation with the retention of the transition metal catalyst in the ionic liquid phase. So far, various common organic reactions employing metal catalysts have been tried in different ionic liquids.

In addition, RTILs are also exploited in various applications, ranging from quartz crystal microbalance sensors (*6*), surfactants for neoteric biphasic systems (*7*), templates for aerogel synthesis (*8*), lubricants for different material contacts (*9*), desulfurization agents (*10*), stationary phases for gas chromatography (*11*), to solvents for nanoparticle production (*12*).

Using RTILs as polymerization media has recently attracted increasing interest (*13*). Many polymerizations, including free radical (*14*), living/controlled free radical (atom transfer radical polymerization (ATRP), reversible addition-fragmentation chain transfer (RAFT) and nitroxide mediated polymerization (NMP)) (*15*), electrochemical (*16*), condensation (*17*), ring-opening metathesis (*18*), block (*19*) and statistical (*20*) polymerizations have been carried out in different RTILs. In addition to the unique properties mentioned above, RTILs can also lead to significant improvements of polymerization rate, and/or yield. In a recent communication (*14*), we described the large increases in both the rate of polymerization and the molecular weights that are obtained in the polymerization of methyl methacrylate (MMA) in [BMIM][PF$_6$] as compared to polymerizations in benzene. In part, these dramatic increases are because of the high viscosity of the polymerization medium. Chain termination is expected to be diminished in viscous systems. Harrison *et al.* (*21*) used the pulse laser polymerization (PLP) technique to polymerize MMA in 1-butyl-3-methylimidazolium hexafluorophosphate ([BMIM][PF$_6$]). They found that both the propagation and termination rates were significantly affected by [BMIM][PF$_6$]. They attributed the increase of the propagation rate to increased polarity of the ionic liquid solution, which lowered the activation energy of propagation through charge-transfer interactions. On the other hand, the termination rate was decreased simply because of the increased viscosity of the polymerization medium. In this paper, we report the conventional free radical polymerisation of styrene and MMA in a wide range of commercially available hydrophobic RTILs (RTILs in this report are all hydrophobic). The polarity and viscosity of RTILs are measured, and a correlation of the polarity and viscosity with the polymerization behavior is attempted.

Experimental

Materials

MMA and styrene (St) (Aldrich, 99%) were passed through an inhibitor-remover column (Aldrich, disposable) and distilled just before use. Benzene (Fisher, ≥99%) was stirred over concentrated sulphuric acid for more than 24 hours and distilled from freshly crushed calcium hydride. Benzoyl peroxide (BPO) (Aldrich, 97%) was recrystallized from chloroform. The common organic solvents: tetrahydrofuran, methanol, ethanol, isopropanol, chloroform, dichloromethane, toluene, benzene, cyclohexane and hexane (Aldrich, HPLC grade) are used as receievd. All RTILs (Merck) were used as received (Table 1): 1-methyl-3-pentylimidazolium tris(pentafluoroethyl)-trisfluorophosphate ([MPIM][(EF$_5$)$_3$(PF$_3$)]), 1-hexyl-3-methylimidazolium tris(pentafluoroethyl) trisfluorophosphate ([HMIM][(EF$_5$)$_3$(PF$_3$)]), trishexyl-(tetradecyl)phosphonium dicyanamide ([H$_3$TDP]DCy), 1-hexyl-3-methyl-imidazolium hexafluorophosphate ([HMIM][PF$_6$]), trishexyl(tetradecyl)-phosphonium bis[oxalato(2-)]-borate ([H$_3$TDP][BOA$_2$]), n-hexyl-pyridinium tetrafluoroborate ([HP][BF$_4$]), trishexyl(tetradecyl)phosphonium bis-(trifluoromethylsulfonyl) imide ([H$_3$TDP][NTf$_2$]), trishexyl(tetradecyl)-phosphonium tris(pentafluoroethyl) trisfluorophosphate ([H$_3$TDP][(EF$_5$)$_3$(PF$_3$)]), 1-butyl-1-

methyl-pyrrolidinium bis-(trifluoromethylsulfonyl)imide ([BMPy][NTf$_2$]), trishexyl(tetradecyl)-phosphoniumchloride ([H$_3$TDP]Cl), 1-hexyl-3-methylimidazolium bis(trifluoromethylsulfonyl)imide ([HMIM][NTf$_2$]) and 1-butyl-3-methyl-imidazolium hexafluorophosphate ([BMIM][PF$_6$]).

Polymerization

All polymerizations were performed under high vacuum to ensure that the systems were oxygen free. The initiator, monomers and reaction media were weighed into ampoules separately. After five cycles of freeze-vacuum-thaw, the ampoules were flame-sealed and kept in a water bath at 70°C for the desired time. The polymerizations were stopped by precipitation into methanol, washing with large amounts of methanol, filtering and drying under high vacuum to constant weight. Conversions were determined gravimetrically.

Instrumentation

Solubilities of RTILs in common organic solvents were tested by mixing RTILs and organic solvents with the same volumes. After vigorously stirring, the systems were then visually checked to see whether they are immiscible, partially miscible or miscible. Size exclusion chromatography (SEC) experiments were carried out at 30 °C using a Polymer Laboratories Knauer 501 HPLC pump and Knauer 2301 RI detector. Two PSS (Polymer Standard Service) columns were used (one 100Å and one linear mixed bead column). The flow rate of the mobile phase (tetrahydrofuran, THF) was 1 mL/min. The ^{1}H and ^{13}C spectra were recorded on a Bruker AMX 400 MHz NMR spectrometer using CDCl$_3$ or CD$_3$COCD$_3$ as solvents and internal references. Differential Scanning Calorimetry (DSC) was carried out on a DSC Q1000 (TA Instruments) in the Polymer Characterization Laboratory at the Chemistry Department (The University of Tennessee at Knoxville) under nitrogen using heating rate 10°C min^{-1}, the second heating circles were used to calculate the glassy transition temperatures. Polarity tests were run on a Perkin-Elmer Instruments, Lamda 35 UV/Vis Spectrometer (Perkin-Elmer). Transition energy, E, of the solvatochromic band, was calculated by the equation: E=28591.44/λ_{max} in kcal/mol or =(hCN$_A$/λ_{max})×106 (h is Planck's constant, C is the speed of light, N$_A$ is Avogadro's number and λ_{max} is the wavelength of maximum absorption; 1 cal is 4.184 J).Viscosities were measured using an AERS-LS #0012005 rheometer (Rheometric Scientific (TA Instruments)): Angular velocity: 7.0498 (rad/s).

Table 1. List of RTILs

RTILs	Abbreviations
1-butyl-3-methyl-imidazolium hexafluorophosphate	[BMIM][PF6]c
1-methyl-3-pentylimidazolium tris(pentafluoroethyl)-trisfluorophosphate	$[MPIM][(EF_5)_3(PF_3)]$
1-hexyl-3-methylimidazolium tris(pentafluoroethyl) trisfluorophosphate	$[HMIM][(EF_5)_3(PF_3)]$
trishexyl-(tetradecyl)phosphonium dicyanamide	$[H_3TDP]DCy$
1-hexyl-3-methyl-imidazolium hexafluorophosphate	$[HMIM][PF_6]$
trishexyl(tetradecyl)-phosphonium bis[oxalato(2-)]-borate	$[H_3TDP][BOA_2]$
n-hexyl-pyridinium tetrafluoroborate	$[HP][BF_4]$
trishexyl(tetradecyl)phosphonium bis-(trifluoromethylsulfonyl)imide	$[H_3TDP][NTf_2]$
trishexyl(tetradecyl)-phosphonium tris(pentafluoroethyl)trisfluorophosphate	$[H_3TDP][(EF_5)_3(PF_3)]$
1-butyl-1-methyl-pyrrolidinium bis-(trifluoromethylsulfonyl)imide	$[BMPy][NTf_2]$
trishexyl(tetradecyl)-phosphonium chloride	$[H_3TDP]Cl$
1-hexyl-3-methylimidazolium bis(trifluoromethylsulfonyl)imide	$[HMIM][NTf_2]$

Results and Discussion

RTILs in this work cover a wide range of cations and anions: typical cations are imidazolium, phosphonium, pyridinium and pyrrolidinium and typical anions include amide, imide, borate, chloride, phosphate and phosphinate (Table 1). Solubilities of the RTILs were tested in common organic solvents and the solubility behaviors of all of the RTILs are similar to that of [BMIM][PF$_6$] (*14*).

Polymerizations of St or MMA were carried out in RTILs using BPO as initator. In order to compare the results, control samples were prepared in benzene under the same conditions. Polymerizations were run for limited periods of time to keep conversions low and avoid complications such as would be expected from the Trommsdorf effect. The characterization data are summarized in Table 2.

Table 2. Characterization Data for Polymerization of MMA and St in Various RTILs or Benzene

Reaction medium	Reaction time (hours)[a]	M_w^b ($\times 10^{-3}$)	M_w/M_n^b	Conv. (%)
MMA polymerization at 70°C				
Benzene[c]	2		No polymer	
Benzene	0.5	91.0	1.85	8.48
[BMIM][PF6][c]	2	129	1.78	4.89
[MPIM][(EF₅)₃(PF₃)]	0.5	451	2.25	29.9
[HMIM][(EF₅)₃(PF₃)]	0.5	450	1.89	29.5
[H₃TDP]DCy	0.5	261	4.43	33.5
[HMIM][PF₆]	0.5	585	2.02	36.6
[H₃TDP][BOA₂]	0.5	385	3.29	23.3
[HP][BF₄]	0.25	707	2.83	38.0
[H₃TDP][NTf₂]	0.5	386	3.28	23.7
[H₃TDP][(EF₅)₃(PF₃)]	0.5	196	2.24	21.9
[BMPy][NTf₂]	0.5	410	1.82	31.5
[H₃TDP]Cl	0.5	236	3.69	0.64
[HMIM][NTf₂]	0.5	436	2.72	26.7
St polymerization at 70°C				
Benzene	1	23.8	1.68	0.33
[MPIM][(EF₅)₃(PF₃)]	1	72.7	3.66	2.75
[HMIM][(EF₅)₃(PF₃)]	1	83.7	3.92	2.75
[H₃TDP]DCy	1	63.0	2.22	trace
[HMIM][PF₆]	1	135	4.45	2.20
[H₃TDP][BOA₂]	1	133	3.32	1.10
[HP][BF₄]	1	162	4.26	3.85
[H₃TDP][NTf₂]	1	112	4.54	2.75
[H₃TDP][(EF₅)₃(PF₃)]	1	84.6	3.14	2.46
[BMPy][NTf₂]	1	137	4.73	2.46
[H₃TDP]Cl	1	99.2	6.26	0.40
[HMIM][NTf₂]	1	70.6	3.36	3.40

[a] All polymerizations were carried out with a ratio [Monomer]/[BPO]/[RTIL] = 1g/20mg/5mL, under high vacuum at 70 °C.
[b] Determined by SEC using THF as eluent and polystyrene standards.
[c] MMA was polymerized thermally without initiators under high vacuum at 70 °C.

In all the cases, except [H₃TDP]Cl, the polymerization of MMA reached higher conversions in RTILs as compared to that in benzene (2-4 times higher). At the same time, the weight average molecular weights obtained in RTILs were much higher (2-8 times higher) and polydispersities (PDIs) were from 1.85 to 4.43 with monomodal molecular weight distributions (MWDs). In most cases, the PMMA was miscible with the RTIL. In [H₃TDP]DCy, shortly after raising the temperature to 70°C, there was phase separation and a solid was formed. The polymer had a broad MWD (M_w/M_n=4.43). For another three RTILs: [HMIM][PF₆], [H₃TDP][BO₂] and [HP][BF₄], gel-formation was observed during polymerization. The low RTIL/monomer ratios may explain this. A similar result was found on polymerization of MMA in [BMIM][PF₆]. Typical SEC traces are shown in Figure 1. In the case of styrene, the polymer precipitated during the reaction due to the insolubility of polystyrene in these RTILs. For polymerizations in RTILs the molecular weights were generally higher with higher yields and broad MWDs. The possible reasons that caused the low conversions of MMA in [H₃TDP]Cl and St in [H₃TDP]Dcy are not clear at the present time.

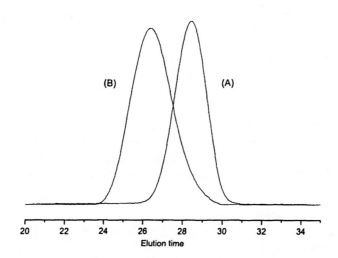

Figure 1. *SEC traces of PMMA (A) polymerized in benzene at 70 °C for 0.5 hour, (B) polymerized in [HMIM][(EF₃)₃(PF₃)] at 70 °C for 0.5 hour.*

8

In the conventional free radical polymerization mechanism, the planarity of propagating free radical chain end (sp^2 hybridization) renders poor control of polymer tacticity because the monomer can attack the chain-end from either side. Stereocontrol may be realized in ionic polymerization (anionic or cationic polymerization) or coordination polymerization (Ziegler-Natta or metallocene catalysized polymerization). Several routes to control the tacticity of free radically produced polymers have been reported: the auxiliary group approach (22), complexation of monomers with Lewis acids (23), chiral initiator or chiral ligand in controlled living free radical polymerization (24), template polymerization or microfluidic channel polymerization (25).

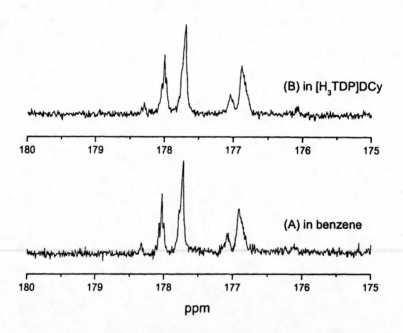

Figure 2. ^{13}C *NMR spectra of PMMAs in (A). benzene and in (B). Trishexyl(tetradecyl)phosphonium dicyanamide ([H₃TDP]Dcy).*

One of our objectives was to determine whether or not conventional free radical polymerization in RTILs can produce polymers with different tacticities as compared to those prepared in common organic solvents. The ^{13}C NMR technique was used to examine the tacticities of all the PMMAs made both in RTILs and in benzene. It is well known that chemical shifts of the carbonyl group of PMMA (~178ppm) are very sensitive to the tacticity. Figure 2 shows a typical set of ^{13}C NMR spectra of the carbonyl region in PMMAs obtained from polymerizations in one of the RTILs ([H₃TDP]Dcy) and in benzene. Our results suggest that the tacticities of PMMAs prepared in RTILs are essentially equivalent to those made in benzene. This is in agreement with our former report comparing results in [BMIM]PF₆ and in benzene (26). Furthermore, DSC traces of PMMAs made by different methods yield similar glass transtion temperatures (T_g) (Figure 3). Since T_g of PMMA is sensitive to tacticity, this gives additional support that free radical polymerization in the RTILs used in the present study yields products that show no significant difference in stereoregularity.

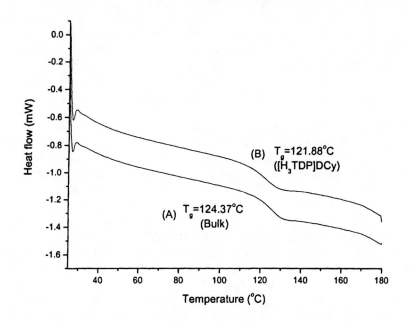

Figure 3.*DSC traces of PMMAs in (A). bulk (M$_w$=400,000) and in (B).
Trishexyl(tetradecyl)phosphonium dicyanamide ([H₃TDP]Dcy) (M$_w$=261,000).*

Although the mechanism of free radical polymerization in RTILs is not completely understood at the present time, some research groups have given tentative explanations. Hong *et al.* (*14*) proposed a "diffusion-controlled termination" mechanism to explain the combination of high molecular weight products and fast reaction rates in RTIL system. The termination rate is diminished due to the relative high viscosity of the medium. This can explain polymerization behavior observed in this research. Harrisson *et al.* (*21*) stated that increased polarity of the medium may favour charge transfer between monomer or radical and ionic liquids. Aki *et al.* (*3*) measured the polarities of various imidazolium and pyridinium ionic liquids and indicated that these ionic liquids are more polar than acetonitrile but less polar than methanol. It was also noted that the preferential radical-ionic liquid complex formation with only one of the ions could introduce electrostatic repulsions between growing chain ends resulting in reduced rates of termination (*21*).

Polarity test results are shown in Table 3. The data obtained in this work are in good agreement with those from the literature (*27*). In the present study, polarities of most of the RTILs were found to lie in the range of small molecular alcohols but other RTILs were found to be as polar as fatty acids. Surprisingly, no clear trend is found between polarity of RTILs or benzene and M_ws or yields of the produced polymers. Attempts to correlate viscosity of the RTILs with polymerization behavior also failed (Table 3).

On the other hand, Strehmel *et al.* (*28*) found that the higher the viscosity, the higher the yields and M_ws of the polymers. This discrepancy may reflect the fact that the RTILs employed in the latter case are composed of cations and anions of similiar structures except for the alkyl substitute length and position on the cation, while either the cations or anions are structurally different in our work. It is still too early to make a definite conclusion on whether or not polarity or viscosity of RTILs have an effect on the polymerization behavior in RTILs. It is our opinion that in correlating viscosity with polymerization behavior it is crucial to keep conversions low, as done in our work, otherwise the Trommsdorf effect can obscure correlations. In the work by Strehmel *et al.*, polymerizations were carried out to high conversions. Likely, the RTILs that are of high viscosity will give polymerization system that exhibit the Trommsdorf effect at lower conversions that for RTILs with lower viscosity. It is well known that the onset of the Trommsdorf effect is accompanied by large increases in polymer molecular weight and rate of polymerization, both caused by diminished termination. Thus, it would be interesting to carried out a study similar to that by Strehmel *et al.* on structurally very similar RTILs but at lower conversions.

Table 3. Correlations between Polarity or Viscosity of RTILs and Characterization Data of Polymers in RTILs

Polarity test[a]					
Ionic liquid	E_{NR} (kJ mol^{-1})	PSt		PMMA	
		M_w^c ($\times 10^{-3}$)	Conv.[d] (%)	M_w^c ($\times 10^{-3}$)	Conv.[d] (%)
[H$_3$TDP][NTf$_2$]	44.33	112	2.75	386	23.7
[H$_3$TDP][(EF$_5$)$_3$(PF$_3$)]	44.62	84.6	2.46	196	21.9
[BMPy][NTf$_2$]	50.99	137	2.64	410	31.5
[MPIM][(EF$_5$)$_3$(PF$_3$)]	51.39	72.7	2.75	451	29.9
[HMIM][(EF$_5$)$_3$(PF$_3$)]	51.60	83.7	2.75	450	29.5
[H$_3$TDP]Dcy	51.65	63.0	trace	261	33.5
[H$_3$TDP]Cl	51.99	99.2	0.40	236	0.64
[H$_3$TDP]BO$_2$	52.05	133	1.10	385	23.3
[HMIM]][NTf$_2$]	52.07	70.6	3.40	436	26.7
[BMIM]PF$_6$	52.17	92.7	5.61	825	22.8
benzene	54.43	23.8	0.33	91.0	8.48

Viscosity test[b]					
Ionic liquid	Viscosity (cP)	PSt		PMMA	
		M_w^c ($\times 10^{-3}$)	Conv.[d] (%)	M_w^c ($\times 10^{-3}$)	Conv.[d] (%)
[H$_3$TDP][(PM$_3$)$_2$P]	955	133	2.33	333	13.3
[MPIM][(EF$_5$)$_3$(PF$_3$)]	234	72.7	2.75	451	29.9
[H$_3$TDP]Dcy	199	63.0	trace	261	33.5
[H$_3$TDP]Cl	137	99.2	0.40	236	0.64
[H$_3$TDP][NTf$_2$]	118	112	2.75	386	23.7
[HP]BF$_4$	109	162	3.85	707	38.0
[H$_3$TDP]BO$_2$	85.1	133	1.10	385	23.3
[HMIM]][NTf$_2$]	79.5	70.6	3.40	436	26.7
[H$_3$TDP][(EF$_5$)$_3$(PF$_3$)]	67.8	84.6	2.46	196	21.9
[HMIM][(EF$_5$)$_3$(PF$_3$)]	56.9	83.7	2.75	450	29.5
[BMPy][NTf$_2$]	27.4	137	2.64	410	31.5

[a] Absorbance range is from 0.5 to 2.0. Transition energy, of the solvatochromic band was calculated by $E = 28591.44/\lambda_{max}$ in kcal/mol.

[b] The viscosity is complex viscosity and was measured at 70°C.

[c] By SEC with polystyrene standars.

[d] Calculated gravimetrically.

The polarity of molecular solvents reflects a combination of many interactions including: H-bonding, p-interaction or van der Waals forces. In RTILs, the situation is even more complicated since both cations and anions may have their own distinct interactions. We used a solvatochromic dye (Nile red) to measure the polarities of RTILs and the results show a similar trend as observed in reference (*29*). Since Nile red is positively solvatochromic, it preferentially interacts with anions of RTILs. The apparent polarities thus deduced maybe not reflect the real values, thus further experiments need to be carried out to address this problem.

MMA can also be thermally polymerized in [BMIM]PF$_6$ without any thermoinitiator at 70°C for 2 hours. The data showed that the polymer produced had M_W of 1.29×10^5 and M_w/M_n of 1.78 with 4.89% conversion. There was no polymer formed in benzene under the same conditions. We hypothesize that the presence of [BMIM]PF$_6$ may aid the spontaneous polymerization of MMA. The catalytic ability of RTILs for organic and bioorganic synthesis has long been recognized and this was discussed in recent review articles (*13*). The catalytic mechanism for polymerization is not understood at this time.

Conclusions

In conclusion, we have shown that the RTILs used in the current study are promising reaction media for free radical polymerization. In most cases, the reactions in RTILs are much faster when compared to those in benzene under the same conditions and lead to polymers with much higher molecular weights. This result indicates that the rapid polymerization rates and high molecular weights observed in previous studies are rather general features of conventional radical polymerization in RTILs. An attempt was made to correlate polarities and viscosities of RTILs with polymerization behavior, however, such a correlation fails to reveal any trends. Thus, additional studies are necessary to elucidate the mechanism that results in the observed accelerations of reactions and the higher molecular weight products.

Acknowledgments

This research was supported by the National Science Foundation (Grant CTS0086874), by the Division of Material Science and Engineering, Office of Basic Energy Sciences, U. S. Department of Energy, under contract No. DE-AC05-00OR22725 with Oak Ridge National Laboratory managed and operated by UT-Battelle, LLC and by Imperial Chemical Industries PLC Strategic Research Fund. We thank Mr. Atul Dahiya and Prof. Simioan Petrovan for viscosity measurements and Prof Bin Hu for polarity measurements. We also thank Prof. Robin Rogers from University of Alabama for some of ionic liquids used in this study.

References

1   Walden, P. *Bull. Acad. Imper. Sci. (St. Petersburg)* **1914**, 1800.
2   (a) Hussey, C. L. *Pure Appl. Chem.* **1988**, *60*, 1763. (b) Seddon, K. R. In *Molten Salt Chemistry*; Mamantov, G.; Marassi, R., Eds.; Reidel Publishing Co.: Dordrecht, The Netherlands, 1987; p 365. (c) Wilkes, J. S.; Levisky, J. A.; Wilson, R. A.; Hussey, C. L. *Inorg. Chem.* **1982**, *21*, 1263. (d) Hussey, C. L. In *Advances in Molten Salts Chemistry*; Mamantov, G.; Mamantov, C., Eds.; Elsevier: New York, 1983; Vol. 5, pp 185-230. (e) Dieter, M. K.; Dymek Jr., C. J.,; Heimer, N. E.; Rovang, J. W.; Wilkes, J. S. *J. Am. Chem. Soc.* **1988**, *110*, 2722. (f) Wilkes, J. S.; Zaworotko, M. J. *J. Chem. Soc., Chem. Commun.* **1992**, 965. (g) Ngo, H. L.; LeCompte, K.; Hargens, L.; McEwen, A. B. *Thermochim. Acta* **2000**, 97, *357-358*.
3   (a) Aki, S. N. V. K.; Brennecke, J. F.; Samanta, A. *Chem. Commun.*, **2001**, 413. (b) Zhang, H.; Hong, K.; Mays, J. W. *Abstract of Paper, ACS 227ᵗʰ National meeting* 2004, Anaheim, CA, USA.
4   Freemantle, M. *C&EN* March 30, 1998, p 32; May 15, 2000, p. 37; January 1, 2001, p 21; May 3, 2004, p 26.
5   Cornils, B.; Herrmann, W. A. In *Applied Homogeneous Catalysis with Organometallic Compounds*; Cornils, B.; Herrmann, W. A., Eds.; Weinheim, New York, 1996, 1167.
6   Liang, C.; Yuan, C.; Warmack, R. J.; Barnes, C. E.; Dai, S. *Anal. Chem.* **2002**, 74, 2172.
7   Merrigan, T. L.; Bates, E. D.; Dorman, S. C.; Davis Jr.,. J. H. *Chem. Commun.* **2000**, 2051.
8   Dai, S.; Ju, Y.; Gao, H.; Lin, J.; Pennycook, S.; Barnes, C. *Chem. Commun.* **2000**, 243.
9   Ye, C.; Liu, W.; Chen, Y.; Yu, L. *Chem. Commun.* **2001**, 2244.
10 Bösmann, A.; Datsevich, L.; Jess, A.; Lauter, A.; Schmitz, C.; Wasserscheid, P. *Chem. Commun.* **2001**, 2494.
11 Anderson, J. L.; Armstrong, D. W. *Anal. Chem.* **2003**, *75*, 4851.
12 Itoh, H.; Naka, K.; Chujo, Y. *J. Am. Chem. Soc.* **2004**, *126*, 3026.
13 (a) Welton, T. *Chem. Rev.* **1999**, 99, 207. (b) Holbrey, J. D.; Seddon, K. R. *Clean Prod. Proc.* **1999**, 1, 223. (c) Wasserscheid, P.; Keim, W. *Angew. Chem. Int. Ed.* **2000**, 39, 3772. (d) Sheldon, R. *Chem.*

14

Commun. **2001**, 2399. (e) Gordon, C. M. *Appl. Catal. A* **2002**, 222, 101.

14 Hong, K., Zhang, H.; Mays, J. W.; Visser, A. E.; Brazel, C. S.; Holbrey, J. D.; Reichert, W. M.; Rogers, R. D. *Chem. Commun.* **2002**, 1368.

15 (a) Carmichael, A. J.; Haddleton, D. M.; Bon, S. A. F.; Seddon, K. R. *Chem. Commun.* **2000**. 1237. (b) Sebastien, P.; Davis, T. P.; Carmichael, A. J.; Haddleton, D. M. *Chem. Commun.* **2002**, 222. (c) Ryan, J.; Aldabbagh, F.; Zetterlund, P. B.; Yamada, B. *Macro. Rapid Commun.* **2004**, 25, 930. (d) Zhang, H.; Hong, K.; Mays, J. W. *Polym. Bull.* **2004**, in press.

16 Naudin, E.; Ho, H. A.; Branchaud, S.; Breau, L.; Belanger, D. *J. Phys. Chem. B* **2002**, 106, 10585.

17 Vygodskii, Y. S.; Lozinskaya, E. I.; Shaplov, A. S. *Macromolecular Rapid Commun.* **2002**, 23, 676.

18 Csihony, S.; Fischmeister, C.; Bruneau, C.; Horvath, I. T.; Dixneuf, P. H. *New J. Chem.* **2002**, 26, 1667.

19 Zhang, H.; Hong, K.; Mays, J. W. *Macromolecules* **2002**, 35, 5738.

20 Zhang, H.; Hong, K.; Jablonsky, M.; Mays, J. W. *Chem. Commun.* **2003**, 1356.

21 (a) Harrisson, S.; Mackenzie, S. R.; Haddleton, D. M. *Chem. Commun.* **2002**, 2850. (b) Harrisson, S.; Mackenzie, S. R.; Haddleton, D. M. *Macromolecules*, **2003**, 36, 5072.

22 Mero, C. L.; Porter, N. A. *J. Org. Chem.* **2000**, 65, 775.

23 Isobe, Y.; Fujioka, D.; Habaue, S.; Okamoto, Y. *J. Am. Chem. Soc.* **2001**, 123, 7180.

24 (a) Puts, R. D.; Sogah, D. Y. *Macromolecules* **1996**, 29, 3323. (b) Haddleton, D. M.; Duncalf, D. J.; Kukulj, D.; Heming, A. M.; Shooter, A. J.; Clark, A. J. *J. Mater. Chem.* **1998**, 8, 1525. (c) Ananchenko, G.; Matyjaszewski, K. *Macromolecules*, **2002**, 35, 8323.

25 (a) Farina, M.; Di Silvestro, G. *J. Chem. Soc., Chem. Commun.* **1976**, 842. (b) Ng, S. M.; Ogino, S.; Aida, T.; Koyano, K. A.; Tatsumi, T. *Macromol. Rapid Commun.* **1997**, 18, 991.

26 Zhang, H.; Bu, L.; Li, M.; Hong, K.; Visser, A. E.; Rogers, R. D.; Mays, J. W. In *Ionic Liquids: Industrial Application to Greener Chemistry*; Rogers, R. D.; Seddon, K. R., Eds.; ACS Symp. Ser. 818, American Chemical Society: Washington, DC, 2001, 114.

27 Carmicheal, A. J.; Seddon, K. R. *J. Phys. Org. Chem.* **2000**, 13, 591.

28 Strehmel, V.; Kraudelt, H.; Wetzel, H.; Görnitz, E.; Laschewsky, A. In *Ionic liquids in polymer systems: Solvents, Additives, and Novel Applications;* Brazel, C. S.; Rogers, R. D., Eds.; ACS Symposium

Series; American Chemical Society: Washington, DC, 2004 Spring, In preparation.

29 (a) Deye, J. F.; Berger, T. A.; Anderson, A. G. *Anal. Chem.* **1990**, 62, 615. (b) Carmicheal, A. J.; Seddon, K. R. *J. Phys. Org. Chem.* **2000**, 13, 591. (c) Dzyuba, S. V.; Bartsch, R. A. *Tetrahedron Lett.* **2002**, 43, 4657.

Chapter 2

Free Radical Polymerization of Methacrylates in Ionic Liquids

Veronika Strehmel[1], André Laschewsky[1,3], Heide Kraudelt[2], Hendrik Wetzel[2], and Eckhard Görnitz[3]

Institutes of [1]Chemistry and [2]Geoecology, University of Potsdam, Karl-Liebknecht-Strasse 24–25, D–14476 Potsdam-Golm, Germany [3]Fraunhofer Institute for Applied Polymer Research, D–14476 Potsdam-Golm, Germany

Imidazolium salts with a modified structure of both, the cation and the anion have been investigated as solvents for polymerization of hydrophobic and hydrophilic monomers. Thermal properties and viscosity of the imidazolium salts are important for their selection as reaction media in polymerization processes. The molecular weight of poly(n-butylmethacrylate) is higher if ionic liquids were used for synthesis in comparison with a conventional solvent. Furthermore, copolymers based on hydrophobic and hydrophilic monomers were successfully synthesized using imidazolium salt as solvent. The amount of the ionic liquid recycled after polymerization without any solvent or by using a traditional solvent depends on both the structure of the polymer and the properties of the imidazolium salt. In case of polymeric sulfobetaines, approximately equimolar amounts of imidazolium salt are bound ionically by the sulfobetaine moiety incorporated in the polyzwitterion.

Introduction

Polymerization processes are strongly influenced by the solvents selected. Low volatile organic solvents have been widely used for polymerization of monomers that are insoluble in water. Substitution of low volatile organic solvents is required for environmental and human health protection. Traditional approaches to substitute volatile solvents are emulsion and suspension polymerization in aqueous solution. A new approach to minimize the amount of low volatile solvents in polymerization processes is the use of ionic liquids (1-3) due to their negligible vapor pressure combined with outstanding solubilization properties. Although ionic liquids have been already widely investigated in synthesis of low molecular organic compounds (4), their use in polymer synthesis has been still in its infancy. In the field of polymer chemistry and engineering, ionic liquids have been investigated as solvents in polymer synthesis (3, 5-18), as plasticizers (19), and in polymer gel electrolytes (20-21). Ionic liquids have been tested in different polymerizations. These are radical (3, 5-16), coordination (3, 17), cationic (18), enzymatic (3), and electrochemical (3) polymerization as well as polycondenzation, and polyaddition (3). Furthermore, atom-transfer radical polymerization (ATRP) (11–13), reverse atom transfer radical polymerization (14-15), and reversible addition-fragmentation chain transfer (RAFT) polymerization (16) were reported as polymerization procedures in ionic liquids. Investigations of free radical polymerization of monomers (1-3, 5-10) show a strong influence of ionic liquids on the properties of the polymers obtained.

Among the monomers, methacrylates have been widely used because of their transparency and their higher glass transition temperature in comparison with acrylates. Furthermore, copolymerization of hydrophobic and hydrophilic monomers is a complicate procedure because of solubility differences between these monomers. Consequently, ionic liquids are interesting reaction surroundings for polymerization of these monomers because of their outstanding solubilization properties for hydrophobic and hydrophilic monomers.

The objective of this study focuses on homopolymerization and copolymerization of butyl methacrylate (1) and 3-(N-[2-methacryloyloxyethyl]-N,N-dimethylammonio)propane sulfonate (2) in ionic liquids. Butyl methacrylate (1) is a hydrophobic apolar model monomer and 3-(N-[2-methacryloyloxyethyl]-N,N-dimethylammonio)propane sulfonate (2) is a model for a highly polar, hydrophilic monomer. Substituted imidazolium tetrafluoroborates (3), imidazolium hexafluorophosphates (4), 1-ethyl-3-methylimidazolium tosylate (5) and 1,3-dimethylimidazolium dimethylphosphate (6) were selected as ionic liquids in order to study their differences in viscosity, and possible effects of the ion structure on polymerization (Table 1). Furthermore, copolymerization of the monomers 1 and 2 in an ionic liquid results in copolymers containing hydrophobic and hydrophilic segments in the same polymer chain. In addition, the polymerization experiments were accompanied by studies of recycling of ionic liquids after polymerization. The recyclability is a crucial aspect to establish ionic liquids in technical processes for economical and ecological reasons.

Experimental

1-Ethyl-3-methylimidazolium tetrafluoroborate (**3a**) and the analogous hexafluorophosphate (**4a**) were purchased from Fluka. 1-Ethyl-3-methylimidazolium tosylate (**5**) and 1,3-dimethylimidazolium dimethylphosphate (**6**) were purchased from Solvent Innovation (Germany). Imidazolium salts **3b – 3f** and **4b – 4f** were synthesized from 1-methylimidazole and the respective alkyl bromide followed by anion metathesis using hexafluorophosphoric acid or tetrafluoroboric acid (22). Imidazolium salts **3g** and **4g** were made in analogy from 1,2-dimethylimidazole and butyl chloride. The monomers (**1**) and (**2**) from Aldrich and the 4,4'-azobisisobutyronitrile from Acros were used as received. Commercial poly(butylmethacrylate) from Fluka was purified by precipitation from toluene solution in methanol.

Solutions of monomers (10 wt%) in ionic liquids were purged with nitrogen to remove dissolved oxygen and then polymerized under nitrogen using 1 mol-% 4,4'-azobisisobutyronitrile at 70°C. The polymers were swollen with the ionic liquid. The poly(butylmethacrylate) (**poly-1**) was isolated by decanting the ionic liquid. Crude **poly-1** was first washed with methanol followed by Soxhlet extraction for 3 hours using methanol to remove traces of the ionic liquid from the polymer. Poly[3-(N-2-(methacryloyloxy)ethyl]-N,N-dimethyl-3-ammonio) propanesulfonate] (**poly-2**) and the copolymers of **1** and **2** were washed with acetonitrile followed by Soxhlet extraction for 8 hours to remove residual ionic liquid remaining in the polymers containing zwitterionic structures. Methanol and acetonitrile were used for purification of the polymers. After the washing process, these solvents were evaporated in vacuum to recycle the ionic liquids. As reference, polymerization experiments of the monomers were carried out in conventional solvents using toluene for polymerization of **1**, water for polymerization of **2**, and acetonitrile containing 4-wt% water in case of copolymerization of **1** with **2**.

NMR spectra were recorded on a Bruker ADVANCE 300 spectrometer. Shear rate dependent viscosities of ionic liquids were determined at 70°C using rheometer with cone plate geometry (Rheometrics DSR 200) in the steady stress modus. Viscosity data are given at a shear rate of 10 s^{-1} where the values become approximately rate independent (Table 1). Thermal transitions of ionic liquids were determined with a Mettler-Toledo DSC 822^e using heating and cooling rates of 5 K/min and 1 K/min, respectively. Thermal stability was investigated under nitrogen with a Mettler-Toledo TGA/SDTA 851^e using a heating rate of 20 K/min. Imidazolium tetrafluoroborates were analyzed using the ion chromatograph Dionex DX100 to detect ionic impurities such as chloride, bromide, and fluoride. The water content in the ionic liquids was determined by Karl-Fischer-Titration using 756 KF Coulometer from Methrom. Molecular weight distributions of poly-1 were measured using size exclusion chromatography (SEC from Waters, column bank: 7.8 mm x 300 mm Waters Styragel HR5, HR4, and HR3; mobile phase: THF, 1 ml/min, detector: RI). Polymethylmethacrylate standards from Polymer Laboratories were used for calibration. The glass transition temperatures of poly-1 were determined with a Mettler-Toledo DSC 822^e using heating and cooling rates of 10 K/min.

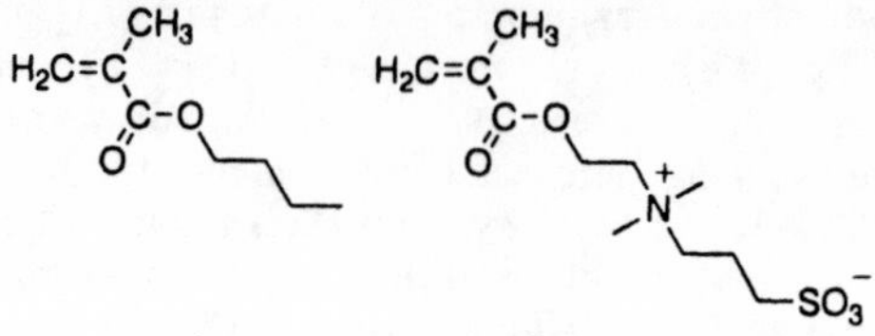

Scheme 1: Structure of monomers

Results & Discussion

The structure of imidazolium salts can be varied broadly via the substitution pattern of the imidazolium cation and by selecting of different anions (Table 1). The substitution pattern of the imidazolium cation was modified by variation of the alkyl chain length (R_1) at nitrogen N-1, and the substituent at C-2 (R_2). The latter is either hydrogen or a methyl group. The methyl substituent at the nitrogen N-3 was the same for all imidazolium salts. Selection of tetrafluoroborate, dimethylphosphate, and tosylate as anions in imidazolium salts results in water-miscible (hydrophilic) ionic liquids. In contrast to this, imidazolium hexafluorophosphates are water-immiscible (hydrophobic) ionic liquids. Nevertheless, all imidazolium salts contain traces of water that are higher in case of imidazolium tetrafluoroborates and 1-ethyl-3-methylimidazolium tosylate (Table 1).

Although all imidazolium tetrafluoroborates contain a certain amount of water, no fluoride was detected by ion chromatography. However, some imidazolium tetrafluoroborates contain traces of chloride originating from the washing process after anion metathesis. Imidazolium hexafluorophosphates were not analyzed by ion chromatography because of their low solubility in water. Structure modification of imidazolium salts results in a variation of their properties. Thermal properties are important for selection of imidazolium salts as new reaction media for polymer synthesis.

Table 1. Impurities (H_2O and Cl^-), thermal properties (glass transition temperature (T_g), recrystallization temperature (T_{cryst}), melting temperature (T_m)), and viscosity (η; measured at 70°C using a rate of $10s^{-1}$) of imidazolium salts of the general structure I

R_1	R_2	X^-		H_2O (wt %)	Cl^- (wt%)	T_g (°C)	T_{cryst} (°C)	T_m (°C)	η (mPa·s)
C_2H_5	H	BF_4^-	3a	0.62	n.d.	-97.5	-67.1	16.4 12.6	10
C_4H_9	H	BF_4^-	3b	1.6	0.0	-87.3	-	-	15
C_5H_{11}	H	BF_4^-	3c	0.3	0.05	-87.4	-	-	22
C_6H_{13}	H	BF_4^-	3d	0.4	0.3	-85.6	-	-	22
C_7H_{15}	H	BF_4^-	3e	0.86	0.0	-86	-	-	25
C_8H_{17}	H	BF_4^-	3f	0.5	0.08	-85.9	-	-	28
C_2H_5	H	PF_6^-	4a	0.01	n.d.	-	-	62.5	20
C_4H_9	H	PF_6^-	4b	0.1	n.d.	-78	-	-12 11	31.7
C_5H_{11}	H	PF_6^-	4c	0.3	n.d.	-76	-	-	40
C_6H_{13}	H	PF_6^-	4d	0.2	n.d.	-74	-	-	35
C_7H_{15}	H	PF_6^-	4e	0.08	n.d.	-72	-	-	44.9
C_8H_{17}	H	PF_6^-	4f	0.1	n.d.	-72	-9	2	48
C_4H_9	CH_3	BF_4^-	3g	0.2	0.0	-73	-18	39	39
C_4H_9	CH_3	PF_6^-	4g	0.2	n.d.	-60	1	16	79.2
C_2H_5	H	H₃C—⟨⟩—SO₃⁻	5	1.9	-	-46	12	53	21
CH_3	H	$H_3CO-\overset{O}{\underset{OCH_3}{P}}-O^-$	6	0.08	-	-64	-	-	11

Thermal Properties and Viscosities of Imidazolium Salts

Temperature stability, melting behavior, and viscosity of ionic liquids are important for their use as solvents in free radical polymerization. These properties are strongly influenced by the structure of the imidazolium salts (Table 1). Thermal stability of imidazolium salts is a limiting factor for their use as reaction sourrounding in polymerization processes (Figure 1). Imidazolium salts, such as 1,3-dimethylimidazolium dimethylphosphate (6) and 1-octyl-3-methylimidazolium tetrafluoroborate (3f) exhibit a small weight loss at about 150°C (Figures 1b and 1c). Therefore, 3f and 6 can be used in polymerization processes only below 150°C. The most stable imidazolium salt investigated in this work is the 1-ethyl-3-methylimidazolium tosylate (5). No weight loss was observed for 5 below 300°C (Figure 1a). Therefore, this ionic liquid can be used for polymerization processes also at high temperatures.

Furthermore, melting behavior of imidazolium salts is important for their selection as new solvents for polymerization processes. As evidenced in Table 1, melting, crystallization, and glass transition are strongly influenced by the structure of both the cation and the anion of the imidazolium salt. 1-Alkyl-3-methylimidazolium tetrafluoroborates and hexafluorophosphates unsubstituted at C-2 and 6 are mainly glass forming materials. Exceptions are 3a, 4b, 4f, methyl substituted imidazolium salts at C-2 (3g and 4g), and 5 showing recrystallizations above the glass transition and melting peaks. 1-Ethyl-3-methylimidazolium hexafluorophosphate (4a) is a crystalline compound. The glass transition and melting temperatures of the imidazolium salts studied show that all imidazolium salts described in this work are liquid at 70°C, which was selected as polymerization temperature.

Viscosity data of imidazolium salts are compared at a shear rate of 10 s^{-1} at 70°C because the values are approximately rate independent. The solvent viscosity is one aspect influencing polymerization processes. As can be seen from the data summarized in Table 1, imidazolium salts with a viscosity lower than 20 mPa·s are 3a, 3b, and 6. The imidazolium salt 4g has the highest viscosity (79 mPa·s at 70°C). All other imidazolium salts possess a viscosity between 20 mPa·s and 50 mPa·s. As a general trend, viscosity of imidazolium salts at 70°C increases with alkyl substituent chain length at the nitrogen N-1, and by introduction of a methyl substituent at C-2 of the imidazolium ring. The latter is more effective concerning viscosity increase. Furthermore, viscosity is slightly effected by the anion. One can observe an increase in melt viscosity as follows:

BF_4^- ~ dimethylphosphate < PF_6^- ~ tosylate (Table 1).

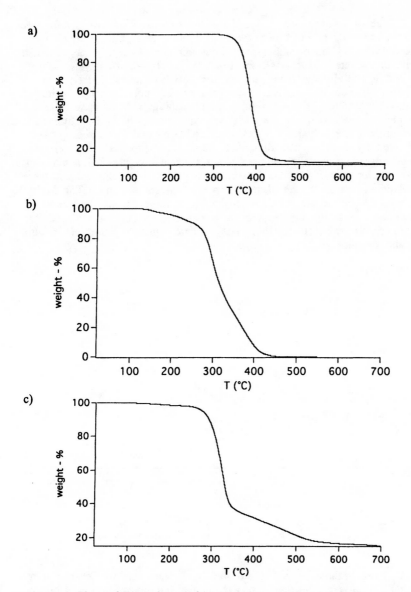

*Figure 1. Thermal Degradation of (a) 1-ethyl-3-methylimidazolium tosylate 5,
(b) 1-octyl-3-methylimidazolium tetrafluoroborate 3f, and (c) 1,3-
dimethylimidazolium dimethylphosphate 6 under nitrogen using a heating rate
of 20 K/min*

Poly(*n*-butylmethacrylate)

Selected imidazolium salts and toluene were used as solvents for free radical polymerization of **1**. The free radical polymerization was initiated using the classical 4,4'-azobisisobutyronitrile as initiator. The polymers obtained were investigated using NMR spectroscopy for structural analysis, GPC to determine the molecular weights and the polydispersities, and DSC to get information about glass transition temperatures. The polymer **poly-1** is free of detectable amounts of ionic liquids after washing with methanol at room temperature as shown by the NMR spectra. Traces of imidazolium salts remaining in the polymer samples undetectable by NMR spectroscopy were removed by Soxhlet extraction after the washing process. Typical GPC chromatograms of various polymer samples depicted in Figures 2, 4, and 6 show higher molecular weight and polydispersity of **poly-1** if imidazolium salts were selected compared to the classical organic solvent toluene. The higher molecular weight of the polymers obtained in ionic liquids may be a result of favored propagation in respect to termination. This was found for polymerization of methylmethacrylate in ionic liquids in comparison with organic solvents (10).

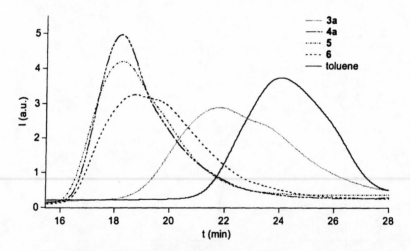

Figure 2: GPC chromatogram of poly-1 synthesized using imidazolium salts as solvents bearing various anions or the traditional toluene

Variation of the anion structure of the imidazolium salts used as solvent in free radical polymerization processes shows that the molecular weight of the polymers obtained is not only influenced by the viscosity of the ionic liquid

(Figure 2). Therefore, structural effects of both the cation and the anion should be discussed as factors influencing the polymerization process as well. Although the viscosity of **3a** and **6** are similar the degree of polymerization ($\overline{X}_n$, Eq. 1) is about 4 times higher if **6** was used as solvent for the polymerization process (Figure 3). Furthermore, the slight increase in the molecular weight of **poly-1** synthesized in **3a** in comparison with using toluene as solvent does not correspond to the large viscosity differences between both media. A similar relation was obtained for imidazolium salts substituted with a methyl group at C-2 as solvents where a similar degree of polymerization was obtained although the viscosity of **4g** is nearly doubled in respect to **3g**. The results discussed demonstrate a stronger effect of the anion on the degree of polymerization as the viscosity.

$$\overline{X}_n = \frac{\overline{M}_n}{M_{monomer}} \qquad (1)$$

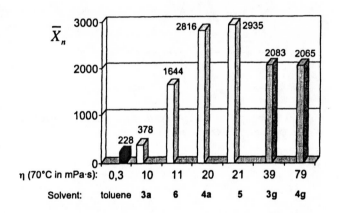

Figure 3: Degree of polymerization ($\overline{X}_n$) of poly-1 synthesized in imidazolium salts or toluene (23) as reaction media depicted as a function of the solvent viscosity (viscosity of toluene was taken from literature (23))

A more clear-cut dependence of the polymerization degree on ionic liquid viscosity was found if the length of the alkyl substituent at the nitrogen N-1 of the imidazolium ring was varied in case of 1-alkyl-3-methylimidazolium tetrafluoroborates. As can be seen from the GPC chromatograms depicted in Figure 4, the maximum of the GPC curve is shifted to higher molecular weight if the length of the alkyl substituent at the nitrogen N-1 results in an increase in viscosity. This may be discussed as one reason for the increase in degree of polymerization (Figure 5). A significantly higher viscosity was measured for

3g. The latter is methyl substituted at C-2 of the imidazolium ring. **Poly-1** synthesized in **3g** exhibits the highest degree of polymerization in case of imidazolium tetrafluoroborates (Figure 5). The systematic increase in viscosity of the imidazolium tetrafluoroborates may be discussed as one reason for the increase in the degree of polymerization in case of **poly-1**.

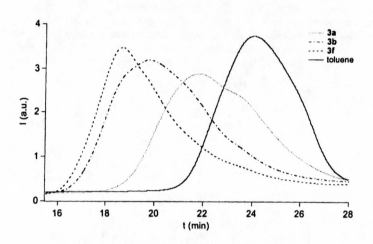

Figure 4: GPC chromatogram of **poly-1** *synthesized using imidazolium tetrafluoroborates or toluene as solvents*

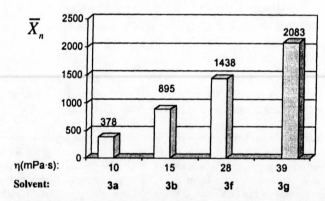

Figure 5: Degree of polymerization ($\overline{X}_n$) of **poly-1** *synthesized using imidazolium tetrafluoroborates as solvents as a function of solvent viscosity*

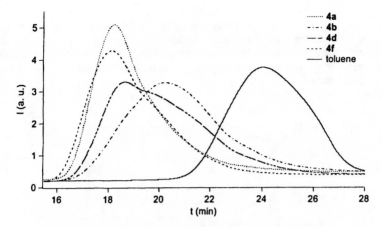

Figure 6: GPC chromatogram of poly-1 synthesized using substituted imidazolium hexafluorophosphates or toluene as solvents

A more complex relation between the molecular weight and the imidazolium salt structure was found if imidazolium hexafluorophosphates of series **4** were used as solvents for polymerization (Figure 6). The degree of polymerization systematically increased if an imidazolium hexafluorophosphate unsubstituted at C-2 was used for polymerization containing an alkyl substitutent at the nitrogen N-1 bearing between 4 and 8 carbon atoms (Figure 7). In contrast to this, the less viscous **4a** results in a higher degree of polymerization of **poly-1** as one would expect particular if viscosity is taken into consideration as a main factor. The degree of polymerization obtained by using **4a** as solvent is nearly the same when **4f** was used as reaction medium for the polymerization experiments although the viscosity of the latter is more than doubled. Furthermore, a lower degree of polymerization of **poly-1** was obtained if **4g** was used as polymerization solvent although the viscosity of **4g** is the highest of all imidazolium salts investigated in this work. Comparing the degree of polymerization of **poly-1** synthesized in **4g** and **3g** shows similar values for both samples (Figures 5 and 7). These results demonstrate the importance of structural factors of the imidazolium salts on order to obtain a high molecular weight of **poly-1** despite the fact that viscosity effects are contributing parameters as well.

The glass transition temperature of **poly-1** depends on the degree of polymerization as depicted in Figure 8. The lowest glass transition temperature was measured for the polymer synthesized in toluene. Furthermore, the glass transition temperature of the commercial **poly-1** that was reprecipitated for purification fits well into this scheme. As expected, a stronger increase in the glass transition temperature is observed in case of a lower degree of polymerization. Only a slight dependence of the glass transition temperature on the degree of polymerization was observed at higher values for the latter. Thus, it can be concluded that ionic liquids are excellent solvents for free radical polymerization of **1** resulting in polymers with higher molecular weight and a higher glass transition temperature. The latter is important for applications of polymers.

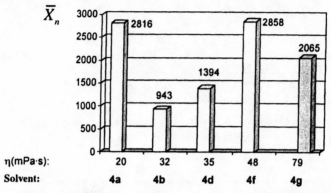

Figure 7: Degree of polymerization ($\overline{X}_n$) of poly-1 synthesized using imidazolium hexafluorophosphates as a function of solvent viscosity

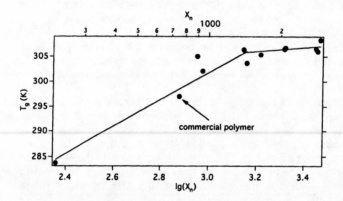

Figure 8: Dependence of the glass transition temperature (T_g) of poly-1 on the degree of polymerization (X_n) for polymers synthesized in toluene or various imidazolium salts as solvents compared with a commercial product

Polymers Containing Zwitterionic Structures

In addition to the apolar monomer **1**, the highly polar zwitterionic monomer **2** was also successfully polymerized in ionic liquids. Different from **1**, the monomer **2** is highly polar due to the sulfobetaine moiety. Therefore, polar solvents are preferred for polymerization of **2** (24). Although this monomer dissolves readily in **4b**, the **poly-2** formed separates from the reaction mixture with progress of the polymerization. This observation agrees with previous reports on poly(sulfobetaines) indicating that solvents are needed that are both polar and protic (24-30). A similar precipitation was found for **poly-1**. Thus, it facilitates isolation of the polymers.

Poly-1 is a hydrophobic polymer that swells with the ionic liquid. DSC measurements of the swollen polymer show both the glass transition temperature of the polymer and the transition temperatures of the ionic liquid because of phase separation of the polymer and the ionic liquid at low temperatures. This indicates only slight interactions between the **poly-1** and the ionic liquid. Typically, **poly-2** exhibits strong interactions with both the cation and the anion of the ionic liquid. This is evident from the NMR spectra of the polymers obtained by polymerization of **2** in **4b** compared to samples prepared in water (Figure 9). The polymer synthesized in **4b** contains approximately a stoichiometric amount of the ionic liquid bound to the zwitterionic moieties even after washing. This observation agrees well with a previous report on the binding of various salts to polyzwitterions in bulk (28). Removing the bound ionic liquid from **poly-2** is not as easy as in case of **poly-1**. Extended Soxhlet extraction is necessary to purify **poly-2**. In addition, traces of nonreacted double bonds were observed in the sample obtained by polymerization in water. However, monomer traces are absent if **4b** was used as solvent.

The successful homopolymerization of **1** and **2** in ionic liquids stimulated the study of the copolymerization of the apolar, hydrophobic monomer **1** and the polar, hydrophilic monomer **2**. Such monomer pairs are difficult to copolymerize in a regular way due to the strongly differing solubility and solvation. The use of ionic liquids with its capacity to dissolve both apolar and polar monomers used is considered as one big advantage for copolymerization of such different monomers. Similar to **poly-2**, the copolymers contain a certain amount of ionic liquid as counter ions. As in the case of the homopolymer, extended Soxhlet extraction was necessary to remove the bound ions quantitatively.

Solubility differences of the copolymer obtained by polymerization in **4b** and the homopolymers let us conclude that copolymers were obtained and not a mixture of two distinct homopolymers. Furthermore, representative signals of both monomer segments were observed in the NMR spectra (Figure 10).

Comparing the NMR spectra of the copolymers synthesized using an excess of **1** in respect to **2** (mole ratio 3:1) in either **4b** or acetonitrile containing a small

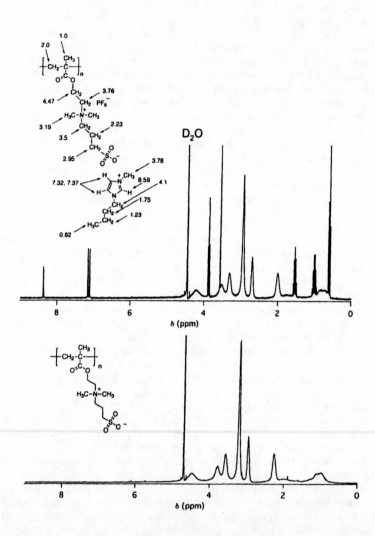

Figure 9: ^1H NMR spectra of poly-2 in D_2O containing 0.1 M NaNO₃ obtained by polymerization of 2 in 4b (top) or acetonitrile containing 4 wt-% water (bottom); assignments for the protons (in ppm) were made based on peak maxima

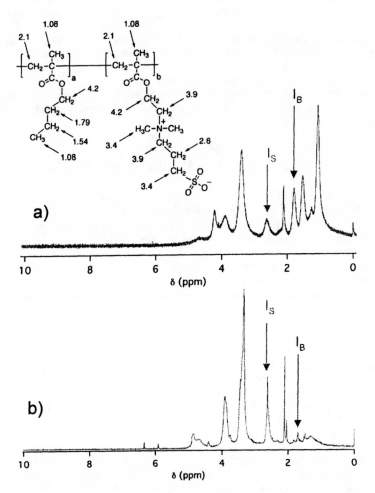

Figure 10: ¹H NMR spectra of copolymers (solvent: CF₃COOD) obtained by copolymerization of 1 with 2 (mole ratio 3:1) using (a) 4b (purification by extended Soxhlet extraction) or using (b) acetonitrile containing 4wt-% water as solvents; assignments for the protons (in ppm) were made based on peak maxima using assignments for butyl methacrylate segments taken from literature (31)

I_S: methylene protons of sulfobetaine segments

I_B: methylene protons of butylmethacrylate segments

amount of water (Figure 10) shows the higher incorporation of the hydrophobic monomer in the copolymer if the ionic liquid **4b** was used for copolymer synthesis ($I_S:I_B = 1:1.4$) although the monomer ratio is the same in both experiments. In case of acetonitrile containing 4wt-% water, the copolymer contains significantly less hydrophobic segments ($I_S:I_B=1:0.2$). A different microscopic surrounding affecting the copolymerization may cause the higher amount of hydrophobic segments in the copolymer synthesized in the ionic liquid. In both cases, the content of segments of **1** is less than one expects from the mole ratio used for copolymer synthesis. Furthermore, unreacted double bonds that were found in the copolymer synthesized in acetonitrile containing 4 wt-% water are absent if the ionic liquid was used during synthesis of the copolymer.

Recycling of Ionic Liquids after Polymerization

The presently high price of the ionic liquids and the demand for environmentally friendly polymerization processes impose the recycling of ionic liquids after the polymerization process. In case of polymerization of **1** more than 90% of the ionic liquid can be recycled and used for further polymerization processes. No detectable monomer and polymer impurities were found by NMR investigation of the recycled ionic liquids. Furthermore, ion chromatography of imidazolium tetrafluoroborates shows no fluoride in the ionic liquids after recycling. The ratio of the ionic liquid recycled by decanting or by using a traditional solvent in a washing process at room temperature depends on the properties of the ionic liquid (Figures 11 and 12). Imidazolium salts substituted with a longer alkyl chain possess a diminished capacity for recycling by decanting as depicted in Figure 11. Imidazolium salts **3f** and **5** are highly viscous or solid at room temperature. Therefore, a traditional solvent is necessary to remove these ionic liquids from the polymer by washing at room temperature. After evaporation of the volatile solvent in vacuum the ionic liquid can be used for further polymerizations. Furthermore, the recycling capability of ionic liquids is also influenced by the anion of the ionic liquid (Figure 12). Imidazolium salts containing tetrafluoroborate, dimethylphosphate, and hexafluorophosphate as anion can be recycled with more than 50% without any traditional solvent. Recycling of **5** needs the use of a conventional solvent to clean the polymer and to recycle the ionic liquid. Evaporation of the low volatile solvent in vacuum results in recycling of the ionic liquid.

Depending on the structure of the ionic liquid, one part of the ionic liquid can be removed from the polymer just by decanting, the other part needs to be recycled by washing with a low volatile solvent and evaporation of this solvent in vacuum. However, isolation and purification of **poly-2** or of copolymers based on **1** and **2** and recycling of the ionic liquid after these polymerization processes require a conventional solvent.

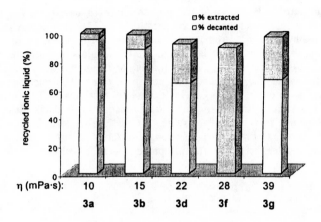

Figure 11: Amount of imidazolium tetrafluoroborates recycled after polymerization of butylmethacrylate by decanting or extraction using methanol at room temperature

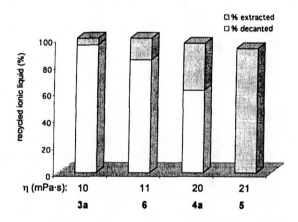

Figure 12: Amount of the imidazolium salts recycled after polymerization of butylmethacrylate by decanting or extraction using methanol at room temperature

Conclusions

Ionic liquids are excellent solvents for free radical polymerization of methacrylates. Homopolymers and copolymers based on the apolar *n*-butylmethacrylate and the polar zwitterionic 3-(*N*-[2-(methacryloyloxy)ethyl]-*N,N*-dimethylammonio propane sulfonate) were successfully synthesized in ionic liquids. Higher molecular weight and higher glass transition temperatures were always obtained using ionic liquids for the polymerization process as compared to toluene. Viscosity and structural effects influence the degree of polymerization of the poly(*n*-butylmethacrylate) synthesized. Polymerization of the sulfobetaine monomer in ionic liquids results in products differing from those obtained in conventional solvents. The use of ionic liquids improves copolymerization of monomers of strongly different polarity. Ionic liquids can be recycled after the polymerization process to fulfill a basic requirement for their economical and ecological use.

References

1. *Ionic Liquids Industrial Applications to Green Chemistry*; ACS Symp. Ser. 818, Rogers, R.D., Seddon, K.R., Eds.; American Chemical Society: Washington, DC, **2002**.
2. *Ionic Liquids as Green* Solvents; ACS Symp. Ser. 856, Rogers, R.D., Seddon, K.R., Kubisa, P., Eds.; American Chemical Society: Washington, DC, **2003**.
3. Kubisa, P. Application of ionic liquids as solvents for polymerization processes. *Progr. Polym. Sci.* **2004**, *29*, 3-12.
4. *Ionic Liquids in Synthesis*; Wasserscheid, P.; Welton, T. Eds.; Wiley-VCH: Weinheim, **2003**.
5. Zhang, H.; Bu, L.; Li, M; Hong, K.; Visser, A.E.; Rogers, R.D.; Mays, J.W. Homopolymerization and Block Copolymer Formation in Room-Temperature Ionic Liquids Using Conventional Free-Radical Initiators. *ACS Symp. Ser. 818*, Rogers, R.D., Seddon, K.R., Eds.; American Chemical Society: Washington, DC, **2002**, 114-123.
6. Benton, M.G.; Brazel, C.S. Effect of room-temperature ionic liquids as replacements for volatile organic solvents in free-radical polymerization. *ACS Symp. Ser. 818*, Rogers, R.D.; Seddon, K.R., Eds.; American Chemical Society: Washington, DC, **2002**, 125-133.

7. Hong, K.; Zhang, H.; Mays, J.W. ; Visser, A.E. ; Brazel, C.S. ; Holbrey, J.D. ; Reichert, W.M. ; Rogers, R.D. Conventional free radical polymerization in room temperature ionic liquids: a green approach to commodity polymers with practical advantages. *Chem. Commun.* **2002**, *13*, 1368-1369.
8. Zhang, H.; Hong, K.; Mays, J.W. Synthesis of Block Copolymers of Styrene and Methyl Methacrylate by Conventional Free Radical Polymerization in Room Temperature Ionic Liquids. *Macromolecules* **2002**, *35(15)*, 5738-5741.
9. Harrisson, S.; Mackenzie, S.R.; Haddleton, D.M. Unprecedent solvent-induced acceleration of free-radical propagation of methyl methacrylate in ionic liquids. *Chem. Commun.* **2002**, *23*, 2850-2851.
10. Harrison, S.; Mackenzie, S.R.; Haddleton, D.M. Pulsed Laser Polymerization in an Ionic Liquid: Strong Solvent Effects on Propagation and Termination of Methyl Methacrylate. *Macromolecules* **2003**, *36*, 5072-5075.
11. Biedron, T.; Kubisa, P. Atom-transfer radical polymerization of acrylates in an ionic liquid. *Macromol. Rap. Commun.* **2001**, *22(15)*, 1237-1242.
12. Biedron, T.; Kubisa, P. Atom transfer radical polymerization of acrylates in an ionic liquid: synthesis of block copolymers. *J. Polym. Sci. Polym. Chem. Ed.* **2002**, *40(16)*, 2799-2809.
13. Biedron, T.; Kubisa, P. Ionic liquids as reaction media for polymerization processes: atom transfer radical polymerization (ATRP) of acrylates in ionic liquids. *Polym. Int.* **2003**, *52(10)*, 1584-1588.
14. Ma, H.; Wan, X.; Chen, X.; Zhou, Q.-F. Reverse atom transfer radical polymerization of methyl methacrylate in room temperature ionic liquids. *J.Polym. Sci. Polym. Chem. Ed.* **2003**, *41(1)*, 143-151.
15. Ma, H.; Wan, X.; Chen, X.; Zhou, Q.-F. Reverse atom transfer radical polymerization of methyl methacrylate in imidazolium ionic liquids. *Polymer* **2003**, *36(14)*, 5072-5075.
16. Perrier, S.; Davis, T.P.; Carmichael, A.J.; Haddleton, D.M. First report of reversible addition-fragmentation chain transfer (RAFT) polymerization in room temperature ionic liquids. *Chem. Commun.* **2002**, *19*, 2226-2227.
17. Shaughnessy, K.H.; Klingshirn, M.A.; P'Pool, S.J.; Holbrey, J.D.; Rogers, R.D. Polar, non-coordinating ionic liquids as solvents for coordination polymerization of olefins. *ACS Symp. Ser.* *856*, **2003**, 300-313.
18. Vijayaraghavan, R.; MacFarlane, D.R. Living cationic polymerization of styrene in an ionic liquid. *Chem. Commun.***2004**, *6*, 700-701.
19. Scott, M.P.; Benton, M.G.; Rahman, M.; Brazel, C.S., Plasticizing Effects of Imidazolium Salts in PMMA: High Temperature Stable Flexible Engineering Materials. ACS Symp. Ser. 856, Rogers, R.D.,

Seddon, K.R., Eds; American Chemical Society: Washington, DC, **2003**, 468-477.

20. De Long, H.C.; Trulove, P.C.; Sutto, T.E. The use of ionic liquids in polymer gel electrolytes. *ACS Symp. Ser. 856*, Rogers, R.D., Seddon, K.R., Eds; American Chemical Society: Washington, DC, **2003**, 478-494.

21. Yoshizawa, M.; Ohno, H. Synthesis of molten salt-type polymer brush and effect of brush structure on the ionic conductivity. *Electrochim. Acta* **2001**, *46*, 1723-1728.

22. Huddleston, J.G.; Visser, A.E.; Reichert, W.M.; Willauer, H.D.; Broker, G.A.; Rogers, R.D. Characterization and comparison of hydrophilic and hydrophobic room temperature ionic liquids incorporating the imidazolium cation. *Green Chemistry* **2001**, *3*, 156-164.

23. Assael, M.J.; Dalaouti, N.K.; Dymond, J.H. The Viscosity of Toluene in the Temperature Range 210 to 370 K. Int. *J. Thermophys.* **2000**, *21(3)*, 291-299.

24. Kudaibergenov, S.E. Recent advances in the study of synthetic polyampholytes in solutions. *Adv. Polym. Sci.* **1999**, *144*, 115-197.

25. Monroy, S.V.M.; Galin, J.C. Poly(sulfobetaines). 1. Synthesis and characterization. *Polymer* **1984**, *25*, 121-128

26. Laschewsky, A.; Zerbe, I. Polymerizable and Polymeric Zwitterionic Surfactants. 1. Synthesis and Bulk Properties. *Polymer* **1991**, *32*, 2070-2080.

27. Anton, P.; Laschewsky, A. Zwitterionic polysoaps with reduced density or surfactant side chain groups. *Makromol. Chem.* **1993**, *194*, 601-624.

28. Köberle, P.; Laschewsky, A. Hydrophobically Modified Zwitterionic Polymers: Synthesis, Bulk Properties, and Miscibility with Inorganic Salts. *Macromolecules* **1994**, *27*, 2165-2173.

29. Bonte, N.; Laschewsky, A.; Vermylen, V. Hybrid Materials Made from Polymeric Betaines and Low Molar Mass Salts. *Macromol. Symp.* **1997**, *117*, 195-206.

30. Favresse, P.; Laschewsky, A. Synthesis and Investigation of New Amphiphilic Poly(carbobetaine)s Made from Diallylammonium Monomers. *Polymer* **2001**, *42*, 2755-2766.

31. Newmark, R.A.; Battiste, J.L.; Koivula, M.N. 2 D NMR of Polymers: Comparison of Some of the Standard Pulse Sequences. *ACS Symp. Ser. 834*, Cheng, H.N.; English, A.D., Eds. American Chemical Society: Washington, DC, **2003**, 179-189.

Chapter 3

Polymer Synthesis in Ionic Liquids: Free Radical Polymerization in Water-Soluble Systems

Carlos Guerrero-Sanchez, Frank Wiesbrock, and Ulrich S. Schubert*

Laboratory of Macromolecular Chemistry and Nanoscience, Eindhoven University of Technology and Dutch Polymer Institute (DPI), P.O. Box 513, 5600 MB Eindhoven, The Netherlands
*Corresponding author: email: u.s.schubert@tue.nl; internet: www.schubert-group.com

During the last years, ionic liquids (ILs) have appeared as serious candidates for the replacement of volatile organic compounds (VOCs) in several chemical processes. Due to the fact that solution polymerization is still one of the most frequently used processes in polymer industry, the replacement of the corresponding solvents with ILs as reaction media seems to be a promising alternative. One of the main problems to be solved before ILs may become real substitutes for organic solvents in this kind of processes is the recovery of the synthesized polymers from the respective reaction media: Since ILs are non-volatile substances, the separation of the polymer from the reaction media turns out to be the bottle-neck and therefore secondary substances must be utilized in order to achieve an effective separation by extraction and/or precipitation. The choice of this secondary substance is one of the most important issues from the ecological point of view, since it will determine whether the process is really environmentally friendly (green chemistry). Taking advantage of the miscibility of some ILs with water and the insolubility of several important commercial polymers in this natural solvent, an alternative for the separation process is addressed. Preliminary results show the feasibility to use water-soluble ILs efficiently as reaction media, which can be explored in a pilot and/or industrial scale for some important industrial polymerization processes. Among the advantages of this alternative approach are: the saving of resources (since ILs can be reused and the polymerization reactions can be performed at lower temperatures than those normally required) and the minimum emission of VOCs into the environment.

Background

One of the main characteristics of ionic liquids (ILs) is their potential to replace volatile organic compounds (VOCs) in several chemical processes,[1-6] including polymer synthesis. The possibility to synthesize polymers in ILs as reaction media has been demonstrated using different polymerization techniques: The polymerization of olefinic monomers in ILs has been widely studied.[1,2,4,6] Conventional free radical polymerization and/or controlled radical polymerization in ILs have also been reported.[7-17] Recently, a living cationic polymerization using an ionic liquid as reaction medium has been published.[18] Other applications of ILs in polymer systems, such as their use as plasticizers,[19] or the synthesis of composite materials[20] and conductive polymers,[7,21,22] have been described as well.

Since solution polymerization processes are required in order to ensure a pumpable mass in the production of several commercial polymers, solution polymerization is still widely used in industry. Many polymers and copolymers are prepared industrially in a continuous process. In that concern, the bulk polymerization method is advantageous to prepare polymers, but comprises problems as the components involved are hard to mix uniformly, and the control of the reaction temperature becomes more difficult in the course of the reaction, concomitant with an increase in viscosity. For this reason, solution polymerization processes are sometimes preferred over bulk processes in order to obtain less viscous media and, consequently, a better control in the process. Important commercial polymers, such as poly(methyl methacrylate) (PMMA), poly(acrylonitrile) (PAN), styrene-acrylonitrile copolymers (SAN) and polyolefins are prepared by solution polymerization.[23-29] Solvents like ethylbenzene or toluene are utilized in large amounts in specific processes to overcome the problems arising from the high viscosity developed during the polymerization. However, this approach has limitations as the yield and the productivity are decreased, and as VOCs are emitted into the environment.

Despite of the fact that some ILs have shown catalytic activity for some chemical reactions[1,2,4,6] and some other intriguing properties, the main interest to use these substances is due to their potential as "green solvents" compared to the well-established and previously unrivaled volatile organic solvents in polymerization processes. In order to develop a more efficient and environmentally friendly solution polymerization process than the actual ones, the use of water-soluble ILs as reaction media is a promising approach. This type of ILs may solve the problem how to separate the polymer from the reaction mixture since some important commercial polymers can be precipitated with the use of water and, consequently, be separated from the aqueous ionic liquid solution by filtration. Considering the mentioned aspects, this work aims at discussing some industrial polymerization processes which require solvents as reaction media in order to screen the feasibility of replacing the corresponding organic solvents by water-soluble ILs. Therefore, the main effort of this contribution is to set up the main criteria for the development of "green" and more efficient industrial polymerization processes utilizing water-soluble ILs.

Actually, there is a considerable number of water-soluble ILs with a promising potential to replace the common VOCs. Table 1 shows a list of some commercially available water-soluble ILs as well as some of their physical properties. The right choice of a water-soluble IL for a specific application in polymerization processes will be constituted mainly by the physical properties of the substance.

Objective

The objective of this study was to propose a methodology for a preliminary evaluation of the ecological advantages of using water-soluble ILs instead of VOCs in industrial solution polymerization processes. The proposed methodology consist of the following steps: 1) Selection of a suitable water-soluble IL (based on the physical properties and the polymerization conditions). 2) Kinetic studies of the polymerization reaction, in order to determine the feasibility of performing the polymerization in the selected IL and to obtain information for the scaling-up to a continuos process. 3) Evaluation of the separation process of the polymer from the reaction media (precipitation of the

Table 1. Water-soluble ionic liquids and some physical properties[30-34]

Water-soluble ionic liquid	Properties
1-Ethyl-2,3-dimethylimidazolium bromide	T_{mp}= 141 °C; T_{fp}= 92 °C; T_{onset}= 322 °C
1-Butyl-2,3-dimethylimidazolium bromide	
1-Hexyl-2,3-dimethylimidazolium bromide	
1-Ethyl-3-methylimidazolium bromide	T_{mp}= 79 °C; T_{fp}= 30 °C; T_{onset}= 311 °C
1-Hexyl-3-methylimidazolium bromide	
1-Ethylpyridinium bromide	
1-Butylpyridinium bromide	
1-Hexylpyridinium bromide	
1-Ethyl-2,3-dimethylimidazolium chloride	T_{mp}= 188 °C; T_{fp}= 103 °C; T_{onset}= 287 °C
1-Butyl-2,3-dimethylimidazolium chloride	
1-Hexyl-2,3-dimethylimidazolium chloride	
1-Ethyl-3-methylimidazolium chloride	T_{mp}= 89 °C; T_{fp}= 33 °C; T_{onset}= 285 °C; $\mu \approx$ 350 cPoise (50 °C)
1-Butyl-3-methylimidazolium chloride	$\mu \approx$ 1000 cPoise (30 °C)
1-Hexyl-3-methylimidazolium chloride	ρ= 1.0593-6.3026x10^{-4}(T-60) g cm^{-3} [T in °C]; $\mu \approx$ 7000 cPoise (30 °C)
1-Octyl-3-methylimidazolium chloride	ρ= 0.9999-3.6033x10^{-4}(T-60) g cm^{-3} [T in °C]; $\mu \approx$ 10000 cPoise (30 °C)
1-Decyl-3-methylimidazolium chloride	
1-Dodecyl-3-methylimidazolium chloride	$T_{mp} \approx$ 52 °C
1-Tetradecyl-3-methylimidazolium chloride	$T_{mp} \approx$ 57 °C
1-Hexadecyl-3-methylimidazolium chloride	$T_{mp} \approx$ 65 °C
1-Octadecyl-3-methylimidazolium chloride	$T_{mp} \approx$ 80 °C
1-Ethylpyridinium chloride	
1-Butylpyridinium chloride	
1-Hexylpyridinium chloride	
1-Ethyl-3-methylimidazolium (L)-lactate	
1-Butyl-3-methylimidazolium (L)-lactate	
1-Butyl-3-methylimidazolium tetrafluoroborate	ρ= 1.1811-7.6229x10^{-4}(T-60) g cm^{-3} [T in °C]; T_{g}= -81 °C; μ= 2.33 Poise; S= 0.0023 S cm^{-1} (30 °C)
1,3-Dimethylimidazolium trifluoromethanesulfonate	T_{mp}= 39 °C

Table I. *Continued*

1-Ethyl-3-methylimidazolium trifluoromethanesulfonate	T_{mp}= -9 °C; μ= 45 cPoise; S= 8.6 mS cm^{-1}; η= 1.4332 ρ= 1.39 g cm^{-3} (20 °C)
1-Butyl-3-methylimidazolium trifluoromethanesulfonate	T_{mp}= 16 °C; μ= 90 cPoise, S= 3.7 mS cm^{-1}; ρ= 1.29 g cm^{-3}; η= 1.438 (20 °C)
1-(2-Methoxyethyl)-3-methylimidazolium trifluoromethanesulfonate	T_{mp}= 27 °C; μ= 74 cPoise; S= 3.6 mS cm^{-1}; η= 1.4428; ρ= 1.364 g cm^{-3}; (20 °C)
1-(2,2,2-Trifluoroethyl)-3-methylimidazolium trifluoromethanesulfonate	T_{mp}= 45 °C
1,3-Diethylimidazolium trifluoromethanesulfonate	T_{mp}= 23 °C; μ= 53 cPoise; S= 7.5 mS cm^{-1}; ρ= 1.33 g cm^{-3}; η= 1.4367 (20 °C)
1-Ethyl-3,4-dimethylimidazolium trifluoromethanesulfonate	T_{mp}= 6 °C; μ= 51 cPoise; S= 6.4 mS cm^{-1}; η= 1.44; ρ= 1.334 g cm^{-3} (20 °C)
1,3-Diethyl-4-methylimidazolium trifluoromethanesulfonate	T_{mp}= 35 °C
1-Ethyl-3-methylimidazolium nonafluorobutanesulfonate	T_{mp}= 28 °C
1,3-Dimethylimidazolium trifluoroacetate	T_{mp}= 52 °C
1-Ethyl-3-methylimidazolium trifluoroacetate	T_{mp}= -14 °C; μ= 35 cPoise; S= 9.6 mS cm^{-1}; η= 1.4405; ρ= 1.285 g cm^{-3} (20 °C)
1-Butyl-3-methylimidazolium trifluoroacetate	μ= 73 cPoise; S= 3.2 mS cm^{-1}; ρ= 1.209 g cm^{-3}; η= 1.4487 (20 °C)
1,3-Diethylimidazolium trifluoroacetate	μ= 43 cPoise; S= 7.4 mS cm^{-1}; ρ= 1.25 g cm^{-3}; η= 1.4431 (20 °C)
1-Butyl-3-ethylimidazolium trifluoroacetate	μ= 89 cPoise; S= 2.5 mS cm^{-1}; ρ= 1.183 g cm^{-3}; η= 1.4441 (20 °C)
1-Butyl-3-methylimidazolium heptafluorobutanoate	μ= 182 cPoise; S= 1 mS cm^{-1}; ρ= 1.333 g cm^{-3}; η= 1.4142 (20 °C)
1-Butylpyridinium trifluorometanesulfonate	
1-Ethyl-3-methylimidazolium tosylate	T_{mp}≈ -15 °C

T_{mp}= melting point, T_{fp}= freezing point, T_{onset}= decomposition temperature according to the thermogravimetric analysis, ρ= density, T_g= glass transition tempertature, μ= dynamic viscosity, S= specific conductivity.

polymer with water, followed by filtration). 4) Recovery of the water-soluble IL from the aqueous solution by distillation in order to recycle the IL in further polymerizations.

Experimental

1-Butyl-3-methyl-imidazolium tetrafluoroborate (BMIM-BF$_4$) and 1-ethyl-3-methyl-imidazolium tosylate (EMIM-TOS) (water-soluble ILs) were purchased from Solvent Innovation GmbH (Germany) and used as received. MMA, BMA and styrene were dried under calcium hydride, distilled and stored under argon at –18 °C prior to use. 2,2'-Azodi(isobutyronitrile) (AIBN) was recrystallized from methanol and stored under argon at –18 °C before use. Kinetic studies of the polymerization of methyl methacrylate (MMA), butyl methacrylate (BMA) and styrene were perfomed in 8 mL vials with magnetic stirring. 0.5 mL of a 1% solution of AIBN in the respective monomer and 0.5 mL of the corresponding water-soluble IL were transferred to each vial. Subsequently, the vials were sealed with a rubber septa and purged with argon for 5 minutes. The vials were placed in a constant temperature oil bath (70 °C). At the transcurred time, each vial was cooled in an ice bath and 0.1 mL of a 1% solution of hydroquinone in tetrahydrofuran was added into the vial. 5 mL of deionized water were added to the obtained suspensions of polymers in each vial in order to complete the precipitation of the polymer. The precipitated polymers were filtered from aqueous solution and dried in a vacuum oven at 50 °C overnight. Conversions were determined by gravimetry. The ILs were recovered from the aqueous solution by distillation under vacuum.

Results & Discussion

The kinetic results of the polymerzation of the selected monomers in the corresponding water-soluble ILs are shown in Figures 1-3. According to Figures 1 and 2, the polymerization rates for MMA and BMA in the water-soluble ILs are higher than those of the bulk polymerization of these monomers. This effect is more pronounced in the case of MMA, which is in full agreement with the results reported in literature[10,15] for the polymerization of MMA in the presence of 1-butyl-3-methylimidazolium hexafluorophosphate (non water-soluble IL). Figure 1 also shows that the typical "gel effect" for bulk polymerization of MMA[35] is stronger for the cases when ILs were used as a reaction media. For the polymerization of styrene (Figure 3), acceleration due to the use of ILs is not evident. In fact, it seems that the use of EMIM-TOS as reaction media causes a small decrease in the polymerization rate of styrene (at least for the reaction conditions used). The observation that the use of ILs as reaction media in free radical polymerizations has almost no effect (or even positive effect) on the polymerization rate (compared to bulk polymerization) is an important aspect. For conventional solution polymerizations, the polymerization rate shows a notable decrease compared to bulk polymerization. The different observations for free radical polymerizations introduce ILs as adequate candidates for the replacement of VOCs in these processes. Although the use of ILs in polymerization processes demands for increasing the volume of the reactors (compared to bulk polymerization), the facts that the polymerization rate is not depleted and that an improved control over the reaction is enabled perfectly counter-balance this change in the volume of the reactors (or the decrease in produced amounts).

Regarding the problem of the separation process, the use of water-soluble ILs allows a satisfying precipitation of PMMA with water for the analyzed cases (Figure 4). After precipitation, the polymers may be separated from the aqueous media by filtration. The ILs can be recovered by distillation and be re-used in further reactions. For the polymerizations of BMA and styrene, the precipitation of the polymers with the only use of water was less effective (but still efficient), especially for low conversion. For this reason, a mixture of methanol:water (20:80%) was used for an effective precipitation in order to obtain reliable kinetic information. It is worth mentioning that, although the monomers and polymers were not fully soluble in the ILs, stable suspensions are developed by slight stirring. Therefore, the polymerization takes place in suspension, which facilitates the precipitation of the polymers with water and renders the use of VOCs unneccesary.

The findings obtained from this work for the polymerization of MMA are especially interesting in terms of the scale-up to a pilot and/or industrial scale. Since the bulk polymerization of MMA is difficult to perform in industrial scale, solution processes are still utilized.[23,24] Thefore the use of water-soluble ILs as reaction media is an alternative for this industrial process. Figure 5 shows a conceptual scheme of a typical industrial continuous process for the polymerization of MMA. Normally, a solution polymerization process is necessary for the industrial production of PMMA in order to ensure a pumpable mass flow in the plant. At the end of the process, solvents (mainly VOCs) must be removed from the reaction mixture in a devolatilization step. This latter procedure is usually carried out by heating the reaction mixture (more than 200°C for a sufficient devolatilization)[23], followed by a strong pressure drop (flash). Moreover, as PMMA has a rather poor thermal stability, this devolatilization step limits the overall yield significantly by partially degrading the polymer back into the monomer.

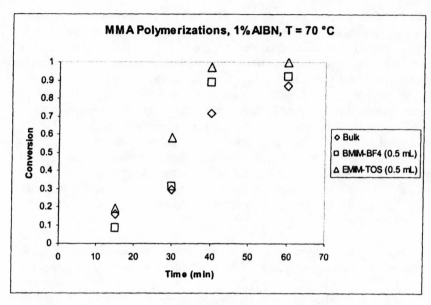

Figure 1. Kinetic measurements for the polymerization of MMA in bulk and in water-soluble ILs.

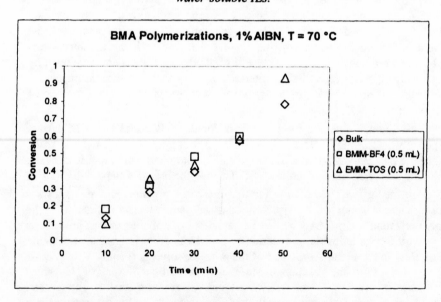

Figure 2. Kinetic measurements for the polymerization of BMA in bulk and in water-soluble ILs.

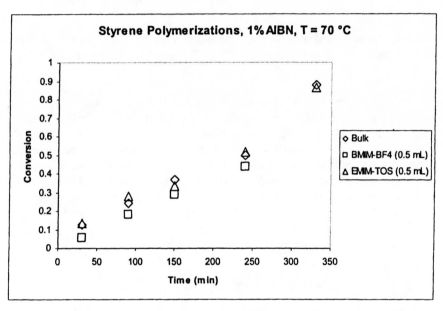

Figure 3. Kinetic measurements for the polymerization of styrene in bulk and in water-soluble ILs.

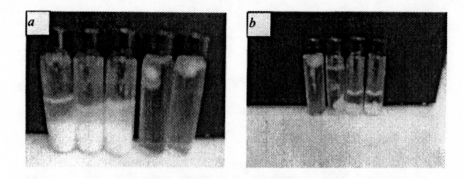

Figure 4. Precipitation of PMMA synthesized in water-soluble ILs with water (different conversion), a) for BMIM-BF₄ and b) for EMIM-TOS.

According to the experimental results obtained in this work, the polymerization of MMA in water-soluble ILs suggests a more efficient process than the commonly used industrial procedures. Figure 6 shows a conceptual proposal, based on the results of this work, for a continuous production of PMMA in an industrial plant using water-soluble ILs as reaction media. Among the advantages of water-soluble ILs in this process, one can find: 1) minium or even no emission of VOCs into the environment, 2) polymerization rates similar to those obtained in bulk polymerization (at least at the explored reaction conditions), 3) saving of energy (the reactions can performed at lower temperature), and 4) the recycling of solvents (ILs and water).

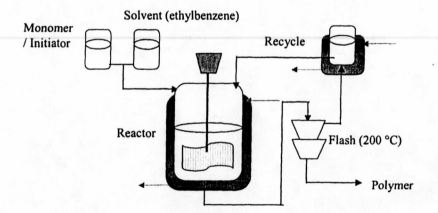

Figure 5. Conceptual scheme for a typical continuous process for the industrial polymerization of MMA.

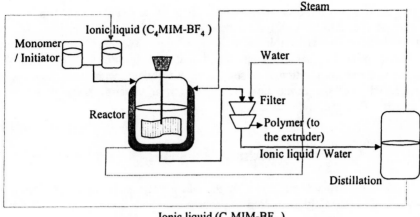

Figure 6. Conceptual proposal for the industrial production of PMMA with the use of water-soluble ILs.

Other industrial polymerization processes may be subjected to the approach presented in this work in order to improve the polymerization reactions. Among these industrial processes are: 1) the production of poly(vinyl chloride) (usually produced by suspension polymerization; the polymer is soluble in 1-butyl-3-methylimidazolium chloride (BMIM-Cl) and regenerable with water), 2) the production of poly(acrylonitrile) (very exothermic reaction in bulk; the polymer is also soluble in BMIM-Cl and regenerable with water), and 3) production of styrene-acrylonitrile copolymers.

Conclusions

In this contribution, a promising route for the development of more efficient and environmentally friendly polymerization processes involving water-soluble ILs has been introduced. Lab scale results show that water-soluble ILs are suitable candidates for the replacement of VOCs in solution polymerization processes. These substances offer several advantages over the traditional VOCs. Future work in this direction will focus on the scale-up of the these experiments to continuously operating pilot plants and/or industrial scale. However, for this emerging field, research is still in its infancy. The design of new processes must be preceded by a full characterization of the physical and chemical properties of water-soluble ILs. Secondly, the number of polymerizations that can be tested in the available water-soluble ILs is enormous. Therefore, methods from combinatorial polymer research may be utilized in order to speed-up the screening and optimization of these systems. Finally, the synthesis of ILs must be improved from the economical point of view in order to make them more attractive for applications in an industrial scale.

Acknowledgments

This research was supported by the Dutch Polymer Institute (DPI) project #401 of the technologic area high-throughput and combinatorial research (HTE).

References

1 Welton, T., *Chem. Rev.*, **1999**; 99, p 2071.

2 Holbrey, J. D.; Seddon, K. R., *Clean Prod Proc.*, **1999**; 1, p 223.

3 Freemantle, M., *Chem. Eng. News*, **2000**; 78(20), p 37.

4 Zhao, D.; Wu, M.; Kuo, Y.; Min, E., *Catalysis Today*, **2002**; 74, p 157.

5 Rogers, R. D.; Seddon, K. R., *Ionic Liquids: Industrial Applications for Green Chemistry*, ACS Ser. 818, Oxford University Press, Oxford, UK, **2002**.

6 Wassercheid, P.; Welton, T., *Ionic Liquids in Synthesis*, Wiley-VCH Verlag GmbH & Co. KgaA, Weinheim, Germany, **2003**.

7 Noda, A.; Watanabe, M., *Electrochimica Acta*, **2000**; 45, p 1265.

8 Carmichael, A. J.; Haddleton, D. M.; Bon, S. A. F.; Seddon K. R., *Chem. Commun.*, **2000**; p 1237.

9 Biedron, T.; Kubisa, P., *Macromol. Rapid Commun.*, **2001**; 22, p 1237.

10 Hong, K.; Zhang, H.; Mays, J. W.; Visser, A. E.; Brazel, C. S.; Holbrey, J. D.; Reichert, W. M.; Rogers, R. D., *Chem. Commun.*, **2002**; p 1368.

11 Zhang, H.; Hong, K.; Mays, J. W., *Macromolecules*, **2002**; 35, p 5738.

12 Klingshirn, M. A.; Broker, G. A.; Holbrey, J. D.; Shaughnessy, K. H.; Rogers, R. D., *Chem. Commun.*, **2002**; p 1394.

13 Perrier, S.; Davis, T. P.; Carmichael, A. J.; Haddleton D. M., *Chem. Commun.*, **2002**; p 2226.

14 Harrison, S.; Mackenzie, S. R.; Haddleton D. M., *Chem. Commun.*, **2002**; p 2850.

15 Harrison, S.; Mackenzie, S. R.; Haddleton D. M., *Macromolecules*, 2003; 36, p 5072.
16 Ma, H.; Wan, X.; Chen, X.; Zhou, Q. F., *Polymer*, 2003; 44, p 5311.
17 Biedron, T.; Kubisa, P., *Polym. Int.*, 2003; 52, p 1584.
18 Vijayaraghavan, R.; MacFarlane, D. R., *Chem. Commun.*, 2004; p 700.
19 Scott, M. P.; Brazel, C. S.; Benton, M. G.; Mays, J. W.; Holbrey, J. D.; Reichert, W. M.; Rogers, R. D., *Chem. Commun.*, 2002; p 1370.
20 Snedden, P.; Cooper, A. I.; Scott, K.; Winterton, N., *Macromolecules*, 2003; 36, p 4549.
21 Yoshizawa, M.; Ogihara, W.; Ohno, H., *Polym. Adv. Technol.*, 2002; 13, p 589.
22 Naudin, E.; Ho, H. A.; Branchaud, S.; Breau, L.; Belanger, D., *J. Phys. Chem. B*, 2002; 106, p 10585.
23 Nising, P.; Zeilmann, T.; Meyer, T., *Chem. Eng. Technol.*, 2003; 26(5), p 599.
24 Higuchi, Y.; Kuwahara, S.; Hieda, S.; Kurokawa, M., *US5719242*, 1998.
25 Uchida, M.; Iwamoto, M.; Nakajima, A.; Takaku, M.; Morita, H.; Kawano, K., *US6060564*, 2000.
26 Gao, X.; Wang, Q.; von Haken Spence, R. E.; Brown, S. J.; Zoricak, P., *US6221985*, 2001.
27 Takami, N.; Uchimura, K.; Yoshioka, K.; Nakazawa, K., *US6403761*, 2002.
28 Tau, L. M.; Swindoll, R. O.; Kao, C.; Jain, P., *US6420516*, 2002.
29 Lee, H. S.; Jang, Y. C., *US6488898*, 2002.
30 Acros Organics, *Ionic Liquids Handbook*, 2003.
31 Bonhôte, P.; Dias, A. P.; Papageorgiou, N.; Kalyanasundaram, K.; Grätzel, M., *Inorg. Chem.*, 1996; 35, p 1168.
32 Ngo, H. L.; LeCompte, K.; Hargens, L.; McEwen, A. B., *Termochim. Act.*, 2000; 357-358, p 97.
33 Bowlas, J. C.; Bruce, D. W.; Seddon, K. R., *Chem. Commun.*, 1996; p 1625.
34 Dullius, J. E. L.; Suarez, P. A. Z.; Einloft, S.; de Souza, R. F.; Dupont, J., *Organometal.*, 1998; 17, p 815.
35 Nising, P.; Meyer, T., *Ind. Eng. Chem. Res.*, 2004; in press.

Chapter 4

Reverse Atom Transfer Radical Polymerization of Methyl Methacrylate in Ionic Liquids

Hongyang Ma and Xinhua Wan*

Department of Polymer Science and Engineering, College of Chemistry and Molecular Engineering, Peking University, Beijing 100871, China

Reverse ATRP of methyl methacrylate (MMA) with AIBN/CuCl$_2$/bipy as an initiating system has been carried out in a number of ionic liquids based on 1,3-disubstituted imidazolium salts. Efforts have been made to adjust the catalyst activity of transition metals, improve the control over macromolecular stereostructure, and recycle the catalyst material by changing reaction conditions and by modifying both the cation and anion structures. All of the ionic liquids employed promote the radical polymerization of MMA to proceed in a controlled manner. The resulting polymethyl methacrylate (PMMA) can be used as a macroinitiator to prepare diblock copolymer with narrow polydispersity via conventional ATRP of second vinyl monomer. In comparison with other reverse ATRPs in bulk or conventional organic solvents, a much smaller amount of catalyst is needed. After a relatively easy removal of polymer and residue monomer, ionic liquids, transition metal and organic ligands can be reused without further purification. As a result, the contamination of the resultant polymer by the catalyst has been largely reduced. The isotacticity of PMMA obtained in ionic liquids containing ester groups or chiral atoms is a little higher than that of PMMA obtained in bulk or other solvents, this implies that there is some solvent effect of ionic liquids on the stereochemistry of radical polymerization.

Background

The past decade has experienced a continued growth of interest and research in the field of ionic liquids as evidenced by the large number of recent publications.[1] Ionic liquids are of considerable importance in chemical technology. They have long been used for a variety of applications such as the production of aluminum, sodium, magnesium, fluorine, fuel cells, nuclear reactors and in chemical synthesis.[2] The recent increased attention and accelerated research activity in ionic liquids are mainly due to their potential use as a new generation of green reaction media.[3,4] The main difference between the properties of ionic liquids and frequently employed organic solvents is their good solubility for metal compounds, non-volatile nature, high viscosity, and recyclablility. It will provide an opportunity in chemical research and industry to resolve many problems and to discover new phenomena which can't be displayed by commonly used organic solvents.

Radical polymerization is a very important commercial process for producing vinyl polymers. It has the advantages of high polymerization reactivity toward many kinds of monomers (especially those with a polar and unprotected functional group), good reproducibility of polymerization results, high tolerance to additives and impurities, the possibility of employing water as solvent and dispersant, and a generally simple overall procedure.[5] The main limitation of radical polymerization is its lack of control over macromolecular structural parameters such as molecular weight (MW), polydispersity, end functionality, chain architecture, composition, and tacticity in comparison with other chain polymerization techniques. Thereafter, much effort has been made to overcome these disadvantages during the last 20 years. Consequently, many new radical polymerization procedures have been developed. The successful techniques include nitroxide mediated polymerization (NMP),[6,7] atom transfer radical polymerization (ATRP),[8-13] and reverse addition-fraction chain transfer (RAFT) polymerization,[14-16] leading to the noticably improved control over some of the key elements of macromolecular structures. Among various available living radical polymerization techniques, ATRP and reverse ATRP are the most extensively studied systems.

In both conventional ATRP and reverse ATRP, the presence of transition-metal-catalyzed reversible atom transfer steps are responsible for the overall control of the process, as shown in Scheme 1.[8-12]

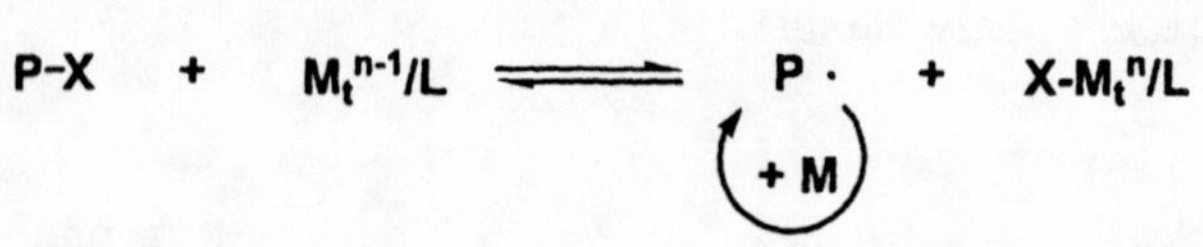

$$P\text{-}X \quad + \quad M_t^{n-1}/L \quad \rightleftharpoons \quad P\cdot \quad + \quad X\text{-}M_t^n/L$$
$$\left(+ M \right)$$

Scheme 1 Mechanism of Atom Transfer Radical Polymerization

The organic halide is employed as the initiator and the low valence transition metal salt as the catalyst. In order to increase the solubility and adjust the redox activity of the catalyst, an organic compound such as bipyridine or the other polydentate N-containing compound is used as a ligand to complex with the metal. The halide first reacts with metal at a lower valence state and donates a halide atom to the salt and itself forms an active radical, which then adds to the double bond of a monomer and initiates the chain reaction. The reaction of the halide with the metal is reversible. The propagation radical can also abstract a halide atom from the higher valence metal salt and produce a domant. In this way, the concentration of active species in living radical polymerization system is kept low and the polymer with predetermined molecular weight and narrow polydispersity can be obtained.

Reverse ATRP is another kind of living radical polymerization technique.[11,12,17,18] The word "reverse" is said relative to conventional ATRP. In conventional ATRP, the organic halide used as an initiator is toxic and the low valent metal is easy to oxidize. To resolve these two problems, reverse ATRP has been developed. Similar to conventional radical polymerization, reverse ATRP employs commonly used radical initiators like AIBN and BPO to induce polymerization and high valence metals to mediate the reversible redox reaction. The mechanism is shown in Scheme 2.

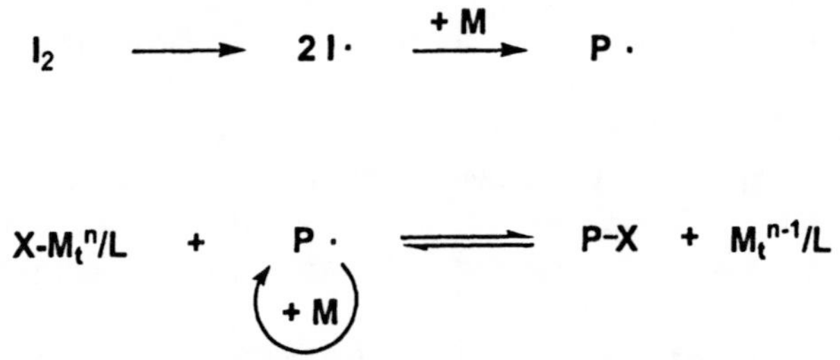

$$I_2 \longrightarrow 2\,I\cdot \xrightarrow{+\,M} P\cdot$$

$$X\text{-}M_t^n/L \;+\; P\cdot \;\rightleftharpoons\; P\text{-}X \;+\; M_t^{n-1}/L$$

$$(+\,M)$$

Scheme 2 Mechanism of reverse Atom Transfer Radical Polymerization

To enable an acceptable reaction rate, high level of catalyst and ligand are required in some conventional ATRP or reverse ATRP.[19,20] Thereafter, the removal and the recycling of catalytic meterials are necessary. Otherwise, the polymer resulted might be contaminated by metal catalyst and easy to age. Also of importance is that the heavy metal contaminant will pollute the environment, especially when mass production is carried out. It is expected that using ionic liquids as polymerization solvents will make the removal and the recycling of the catalylitc materials easier because of their good solubility of transition metals.

Another large limitation of radical polymerization is that it is not easy to control the stereostructure of the obtained polymer if not impossible.[21] It has been described that the tacticity of the radical initiated vinyl polymer can be improved by employing bulky monomer,[22] addition of Lewis acid,[23] or using polar solvents like fluoroalcohol.[24] Ionic liquids are known as "designer's solvent". Their properties can be easily tailored by changing the structure of the cation, anion, or both. We feel that taking full advantage of novel chemistry involved in ionic liquids can not only reduce chemical wastes greatly but can also achieve high level control over the architecture and properties of polymers in radical polymerization of vinyl monomers. This chapter surveys reverse ATRP in ionic liquids, which has been approached recently in our laboratory. Published results are reviewed critically and complemented by a preliminary report of work under way.

Experimental

All the ionic liquids were prepared according to the literature procedure except noted.[25] As an example, the preparation of the 1-butyl-3-methylimidazolium hexafluorophosphate ([bmim][PF$_6$]) is shown as following.

To a dry glass tube, 8.20 g (0.10 mol) of 1-methylimidazole and 11.40 ml (0.13 mol) of 1-chlorobutane was added. After three freeze-pump-thaw cycles, the tube was sealed under vacuum and placed into an oil bath thermostated at 70 ℃. The reaction continued for three days and a pale-yellow viscous liquid appeared. Upon cooling to 0 ℃, a white solid, 1-butyl-3-methylimidazolium chloride, was obtained. The solid was washed three times with ethyl acetate and then dissolved in DI water. To the aqueous solution, which was cooled by ice-bath, 19 ml of hexafluorophosphoric acid (65% water solution, 0.13 mol) was added slowly. After stirring over night, an orange viscous liquid was obtained. The liquid was diluted with dichloromethane, and washed with DI water until it became neutral. The resulted ionic liquid was dried under vacuum at 70 ℃ after evaporation of dichloromethane.

A typical procedure for reverse ATRP of MMA in ionic liquid is described below.

A dry glass polymerization tube was charged with CuCl$_2$, MMA, bipyridine (bipy), 2,2'-azobisisobutyronitrile (AIBN) and ionic liquid. The mixture was degassed by three freeze-pump-thaw cycles and sealed under vacuum. The tube was placed in a water bath at the desired temperature maintained by a thermostat bath. After an expected time, the tube was placed into an ice bath to stop the reaction. The reaction mixture was diluted with THF and added dropwise into methanol. After filtration and washing three times by methanol, followed by drying under infrared light for 24 hours, PMMA was obtained. The conversion of monomer was determined gravimetrically.

The number-averaged molecular weights and molecular weight distributions of PMMA were measured on a Waters 2410 GPC instrument with a set of HT2+HT3+HT4 μ-Styragel columns using THF as eluent (1.0 ml/min) at 35 °C. Calibration was made with a series of standard polystyrenes. ^{1}H NMR spectra were taken at 25 °C on a Bruker ARX400 NMR spectrometer using chloroform-d as solvent and tetramethylsilane (TMS) as an internal reference.

Results & Discussion

Reverse ATRP in [bmim][PF$_6$]

Ionic liquids with 1,3-dialkylimidazolium as the cation and hexafluorophosphate as the anion have proved to be good solvents for radical polymerization of vinyl monomers.[26-31] Studies on Cu^I mediated ATRP of MMA showed that the polymerization rate was enhanced and polymers with narrow dispersity were obtained.[32] Furthermore, the reaction temperature could be lower than that in other solvents. Another interesting thing thereof was that the contamination of polymer by metal catalyst could be largely avoided.

Sarbu and Matyjaszewski have studied ATRP of MMA in the presence of ionic liquids with different anions.[33] They found that only catalytic amount of ionic liquids could achieve the living polymerization. When Iron(II) was used as a catalyst, no organic ligand was needed no matter what kind of anion was employed. However, organic ligand was necessary when ionic liquids with halide and carbonate anions were used in the copper(I) catalyzed ATRP of MMA, but not for ionic liquids with phosphonate as the anion. In addition, the catalytic system was recycled and reused after removal of the polymer and residual monomer.

The solubility of the acrylates in [bmim][PF$_6$] depends on the substituent.[34] When CuBr complexed with pentamethyldiethylenetriamine was used as a catalyst for ATRP, methyl acrylate gave a homogeneous polymerization system and the polymers obtained had molecular weights close to calculated values and had narrow molecular weight distributions. Although higher acrylates gave heterogeneous polymerizations with the catalyst present in the ionic liquid phase, which deviated from living behaviour, ATRP of butyl acrylate in ionic liquids followed by addition of methyl acrylate allowed the clean synthesis of AB block copolymers by one-pot sequential polymerization even if the first stage was carried out to complete conversion of butyl acrylate.[35]

Xi and coworkers studied atom transfer radical polymerization of N-substituted maleimides with styrene in [bmim][PF$_6$].[36,37] They used dendritic polyarylether 2-bromoisobutyrates of different generations as macroinitiators and CuBr/pentamethyldiethylenetriamine as a catalyst. The resulted dendritic-linear block copolymers had well-defined molecular weight, low polydispersity, and could be used as a macroinitiator for chain-extension reaction. In addition, the polymerization in ionic liquids had a much stronger tendency for alternation when compared to the polymerization in anisole. The tendency for alternation decreased in the order N-phenylmaleimide (PhMI) > N-butylmaleimide (NBMI) > N-cyclohexylmaleimide (ChMI) in [bmim][PF$_6$] and PhMI > ChMI > NBMI in anisole.

Reverse ATRPs of MMA using AIBN/CuCl$_2$/bipyridine (bipy) as the initiating system were carried out at 60, 70, and 90 ˚C, respectively, in [bmim][PF$_6$].[38] The molar ratio of [MMA]/[AIBN]/[CuCl$_2$]/[bipy] was 100:1:2:6. Throughout the polymerization, the initiator, transition metal salt, ligand, momoner, and polymer were soluble in [bmim][PF$_6$], and the reaction mixture remained homogeneous. The polymerization results were summarized in Table 1.

Although it was reported that the heterogeneous AIBN/CuCl$_2$/bipy initiating system could not promote radical polymerization of methyl acrylate and MMA in bulk to proceed in a controlled manner[17,18], living behavior in [bmim][PF$_6$] with the same catalytic system was observed. This may be due to the good solubility of the catalyst complex in the ionic liquid. Thus, the concentration of Cu^{II} was much higher than that in a heterogeneous system when the same amount of CuCl$_2$ was added.

The higher visicosity of [bmim][PF$_6$] compared to bulk MMA and other organic solvent has exerted a profound influence on initiator efficiency, which was defined as f = $M_{n,th}/M_{n,GPC}$. It can be seen from Table 1 that the molecular weights measured by GPC, $M_{n,GPC}$, were much higher than the calculated values, $M_{n,th}$, according to $M_{n,th} = 100 \times (\triangle [MMA]/2[AIBN]_0) \times$ conversion, implying that the initiator efficiency of AIBN in the ionic liquid was lower than that in conventional organic solvents. Sarbu and Matyjaszewski have reported a similar behavior.[33] They ascribed the low f value to the relatively low concentration of

Table 1. Reverse ATRP of MMA in [bmim][PF₆] at different temperatures.[a]

Entry	Temp., °C	Time, hour	Conv., %	$M_{n, th}$ $\times 10^{-3b}$, Dalton	$M_{n, GPC}$ $\times 10^{-4c}$, Dalton	M_w/M_n	f^d
1		11.6	12.56	0.63	0.65	1.19	9.69
2		14.3	22.66	1.13	0.86	1.19	13.14
3	60	22.5	47.53	2.38	1.88	1.08	12.66
4		24.4	56.67	2.83	2.03	1.08	13.94
5		26.3	66.67	3.33	2.23	1.08	14.93
6		30.8	87.33	4.37	2.62	1.08	16.68
7		6.0	26.54	1.33	1.26	1.11	10.55
8		7.5	44.9	2.25	1.53	1.09	14.71
9	70	8.0	51.91	2.60	1.53	1.09	16.99
10		9.5	64.46	3.22	1.76	1.08	18.3
11		11.7	79.49	3.97	1.77	1.09	22.43
12		12.5	89.41	4.47	2.03	1.08	22.02
13		0.5	20.34	1.02	0.44	1.18	23.18
14		0.8	45.73	2.29	0.81	1.17	28.27
15	90	1.3	63.57	3.18	1.43	1.08	22.24
16		1.9	74.11	3.71	1.60	1.08	23.19
17		2.3	77.08	3.85	1.62	1.08	23.77
18		2.6	81.04	4.05	1.65	1.09	24.55

[a] MMA = 0.50 g; [bmim][PF₆] = 1.50 g; [MMA]:[AIBN]:[CuCl₂]:[bipy] = 100:1:2:6.

[b] Calculated based on $M_{n, th} = 100 \times (\Delta[MMA]/2[AIBN]_0) \times$ conv.

[c] By GPC with polystyrene standards.

[d] Initiating efficiency: $f = M_{n, th}/M_{n, GPC} \times 100$ (%).

Source: Reproduced with permission from reference 38. Copyright 2003 John Wiley and Sons.

the catalyst in the organic phase and the very high concentration of the catalyst in the ionic phase. However, the concentration difference of the catalyst in two separate phases could not account for the low initiator efficiency in the above experiments because all the reagents were soluble in [bmim][PF₆] and a homogeneous solution was formed. The low initiating efficiency was most likely due to the cage effect of [bmim][PF₆] molecules. Because of the intense confinement of the highly viscous solvent, some side reactions of the primary radicals might have occurred, and as a result, the concentration of the initiator radicals was depleted.

The chemical structure of the obtained PMMA was characterized by 400-MHz ¹H NMR spectroscopy as

$$CH_3-\underset{\underset{CN}{|}}{\overset{\overset{CH_3}{|}}{C}}\!\!-\!\!\left(\!CH_2-\underset{\underset{\underset{O}{\diagdown}\,OCH_3}{C}}{\overset{\overset{CH_3}{|}}{C}}\!\right)_{\!n}\!\!CH_2-\underset{\underset{COOCH_3}{|}}{\overset{\overset{CH_3}{|}}{C}}\!\!-\!\!Cl$$

To prove the chemical structure of the resultant polymer further, the chain extension and block copolymerization were made by conventional ATRP of MMA and styrene, respectively, in chlorobenzene, with chlorine-ended PMMA as a macroinitiator, CuCl as a catalyst, and bipyridine as a complexing ligand. The GPC curves of the resultant polymer and copolymer shifted to higher molecular weights obviously, implying the well-defined end-group.

One advantage of reverse ATRP of MMA in [bmim][PF$_6$] over traditional methods is that a much smaller amount of catalyst is needed to effectively mediate the polymerization process. To investigate the effect of the CuCl$_2$ content on the reverse ATRP of MMA in [bmim][PF$_6$], we employed a variety of initiating systems, AIBN/CuCl$_2$/bipy, with varied [CuCl$_2$]/[AIBN] ratios of 0-8, while keeping constant [MMA]/[AIBN] at 100 and [bipy]/[CuCl$_2$] at 3. The results were summarized in Table 2. In the absence of CuCl$_2$, the molecular weight distribution of the obtained PMMA was broad, suggesting an uncontrolled process. However, the polymerizations became well controlled when only 0.125 equiv of CuCl$_2$ versus AIBN was added. All the polymers had

Table 2. Reverse ATRP of MMA in [bmim][PF$_6$] at different [AIBN]/[CuCl$_2$] ratio.[a]

Entry	[AIBN]:[CuCl$_2$]	$M_n \times 10^{-4}$	M_w/M_n	Yield (%)
1	1:0	18.65	1.54	99.00
2[b]	1:0	17.98	1.55	99.00
3	1:0.125	3.17	1.10	97.89
4	1:0.25	2.62	1.09	92.05
5	1:0.5	2.46	1.09	90.67
6	1:1	1.56	1.06	58.32
7	1:2	1.60	1.07	53.77
8	1:4	1.51	1.07	41.81
9	1:5	1.38	1.07	34.91

[a] MMA = 0.50 g; [bmim][PF$_6$] = 1.50 g; [MMA]:[AIBN] = 100:1; [CuCl$_2$]:[bipy] = 1:3; temperature = 70 □; reaction time = 8 h.

[b] Sample was prepared by conventional free radical polymerization but with 5% bipyridine added.

Source: Reproduced with permission from reference 38. Copyright 2003 John Wiley and Sons.

relatively narrow M_w/M_n values (less than 1.10). The polymerization rate increased with an decrease in the [CuCl$_2$]/[AIBN] ratio, as evidenced by the increased monomer conversion at the same polymerization time. This effect was due to more radicals being generated by the decomposition of AIBN.

It is interesting to ask why the amount of CuCl$_2$ needed to gain control over the reverse ATRP of MMA was less in an ionic liquid than in bulk or conventional organic solvents. According to the mechanism of reverse ATRP proposed by Matyjaszewski and coworkers,[11,12] 1 molar equivalent of Cu^{II} species versus growing radicals was required to promote a controlled polymerization. However, the [AIBN]/[CuCl$_2$] ratio here was as large as 8 when a well-controlled process was observed. The main reason might be the low efficiency of AIBN in [bmim][PF$_6$], in addition to the good solubility of Cu^{II} species in the ionic liquid.

The probability of recycling [bmim][PF$_6$] and catalytic system was evaluated. When polymerization was stopped, the reaction mixture was diluted with tetrahydrofuran (THF) and added slowly into a large amount of methanol. The precipitated PMMA was isolated by filtration. After the evaporation of the methanol, THF, and residue monomer under high vacuum, AIBN and MMA were introduced into recovered [bmim][PF$_6$] containing recycled CuCl$_2$ and bipyridine. The reverse ATRP was performed again. The Mn values increased linearly with conversion, and the polydispersities remained relatively narrow; this indicated a well-controlled polymerization process, although no new catalyst and ligand were added.

Reverse ATRP in [bmim][BF$_4$]

1,3-Disubstituted imidazolium tetrafluoroborates are also frequently used as reaction media.[39] Compared to ionic liquids containing PF$_6^-$ anion, they have higher solvating power for transition metal salts.[40]

Reverse ATRP of MMA was carried out at 80 °C in 1-butyl-3-methylimidazolium tetrafluoroborate ([C$_4$mim][BF$_4$]) and 1-dodecyl-3-methylimidazolium tetrafluroborate ([C$_{12}$mim][BF$_4$]), respectively, with AIBN/CuCl$_2$/bipy as an initiating system.[41] All reagents including initiator, transition metal salts, organic ligand and monomer were soluble in both ionic liquids. But PMMA with high molecular weight was not readily soluble in [C$_4$mim][BF$_4$]. The reaction mixture in [C$_4$mim][BF$_4$] kept homogeneous until the monomer conversion was 9.8% and M$_n$ was 4.68 × 10^3 dalton. As polymerization proceeded further, PMMA precipitated and led to an ill-controlled radical reaction. The solubility of PMMA in [C$_{12}$mim][BF$_4$], which has a long dodecyl group, was much higher than that in [C₄mim][BF₄], indicating a strong dependence of solvating power of ionic liquids upon the length of the substituted groups of the cations. The reaction mixture remained homogeneous in [C₁₂mim][BF₄] at 80 ℃ until the monomer conversion reached 80%. Although the polymer precipitated when the monomer conversion exceeded 80%, the polymerization still proceeded in a well-controlled manner until the monomer conversion got to 92%. However, when the monomer conversion was above 92%, the polymerization became ill-controlled. Similar to reverse ATRP of MMA in [bmim][PF₆], the initiating efficiency f was also lower in [C₁₂mim][BF₄] than that in conventional organic solvent.

The kinetic studies proved the living nature of reverse ATRP of MMA in [C₁₂mim][BF₄]. Figure 1(left) shows a linear relationship between ln ([M]₀/[M]) and reaction time, implying the concentration of the propagation radicals was constant during the polymerization. Furthermore, the resultant polymer has a relatively narrow polydispersity and the number-averaged molecular weight (M_n) increased linearly with monomer conversion, as shown in Figure 1 (right).

Meanwhile, the recycle and reuse of [C₁₂mim][BF₄] was performed. As shown in Table 3, the molecular weights, polydispersities, and yields of the polymers obtained in recycled [C₁₂mim][BF₄] were almost unchanged, while other conditions were kept identical.

The stereochemistry of reverse ATRP of MMA in [C₁₂mim][BF₄] was also approached. However, the stereoregularity of the resultant polymers was not obviously different from that in conventional solvent and bulk.[42]

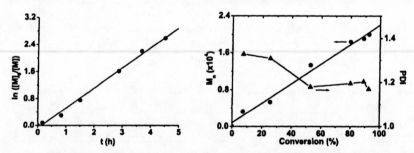

Figure 1. Dependence of ln([M]₀/[M]) on the polymerization time (left) and M_n, M_w/M_n versus conversion (right). Polymerization conditions: [MMA]:[AIBN]:[CuCl₂]:[bipy] = 100:1:2:6, [MMA]: [C₁₂mim][BF₄] = 1:3 (w/w), T = 80 ℃.

(Reproduced with permission from reference 41. Copyright 2003 Elsevier.)

Reverse ATRP in Ionic Liquids with Ester Group

In the previous two sections, reverse ATRP of MMA in $[C_4mim][PF_6]$ and $[C_nmim][BF_4]$ (n=4,12) has been described. Due to solvent effect, some unique characteristics have been exhibited. Considering the ionic liquids as designer solvent,[43] it is important and interesting to develop ionic liquids with multi-functions. The carbonyl group is able to coordinate with some transitional metal under certain conditions. The introduction of ester groups in the molecules of ionic liquids might endow them with some functions.

We have synthesized a series of ionic liquids containing ester group as shown below:[42]

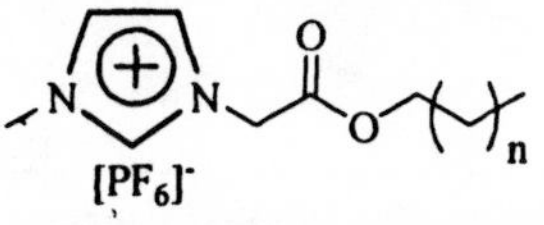

$[PF_6]^-$

n = 0, 2, 4, 6, 8, 10, 12, 14, 16

Reverse ATRP of MMA in IL-1 (n = 0) and IL-2 (n = 2) was carried out and the results were summarized in Table 4. From Table 4, it is evident that all the polymers obtained have well-defined structure, as indicated by narrow polydispersity and the fact that $M_{n,GPC}$ is close to $M_{n,NMR}$. In addition, the isotacticities of the polymers obtained in ionic liquids were a little higher than those in toluene. The reason is not very clear but surely partly because of the introduction of the carbonyl group in ionic liquids. Encourged by this small increase in isotacticity of PMMA, we have designed and prepared several chiral ionic liquids and tried to study stereochemistry of radical polymerization in an asymmetrical environment.

Reverse ATRP in Chiral Ionic Liquids

Chiral ionic liquids, which contain chiral atoms, have received attention because of their potential and unique application in asymmetric synthesis and chiral separation.[44-52] The goal of our work is to investigate the possibility of using them as auxiliary solvents to increase the tacticity of vinyl monomer obtained by free radical polymerization.[53] We synthesized four chiral ionic liquids, 1-((-)-menthoxycarbonylmethylene)-3-methylimidazolium hexafluorophosphate (CIL-1), 1-((-)-menthoxycarbonylmethylene)-3-hexadecylimidazolium hexafluorophosphate (CIL-2), 1-((-)-ethoxycarbonylmethylmethoxymethylene)-3-methylimidazolium hexafluorophosphate (CIL-3), and 1-((-)-ethoxycarbonylmethylmethoxymethylene)-3-hexadecylimidazolium hexafluorophosphate (CIL-4).

Table 3. Results of reverse ATRP of MMA in recovered [C$_{12}$mim][BF$_4$]. [a]

Re-used times	Yield (%)	M$_n$×10^{-4} (Dalton)	M$_w$/M$_n$
1	53.9	1.29	1.14
2	42.5	1.12	1.10
3	42.4	1.23	1.12
4	53.5	1.20	1.15

[a] [MMA]/[AIBN]/[CuCl$_2$]/[bipy] = 100/1/2/6 (molar ratio), [MMA]:[ionic liquid] = 1:3 (w/w), T = 80 ℃, time = 3.7 h.

Source: Reproduced with permission from reference 41. Copyright 2003 Elsevier.

Table 4 Reverse ATRP of MMA in ionic liquids IL-1 and IL-2 [a]

Entry	Solvent	M$_{n,GPC}$ ×10^{-4}, Dalton	M$_{n,NMR}$ ×10^{-4}, Dalton	PDI	Tacticity			Yield , %
					mm	mr	rr	
1[b]	toluene	3.55	-	1.62	5.50	33.20	61.30	67.30
2	IL-1	1.41	2.22	1.16	7.15	36.57	56.28	79.88
3	IL-2	1.23	1.68	1.12	7.92	37.51	54.57	69.38

[a] [MMA]:[AIBN]:[CuCl$_2$]:[bipy] = 100:1:2:6, [MMA]/[ILs] = 1:3 (w/w), T = 90 ℃, time = 24 h.

[b] Free radical polymerization in toluene at 60 ℃ for 24 h.

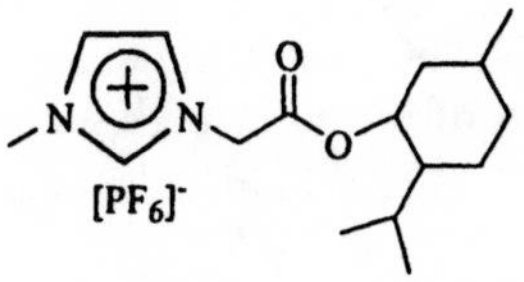

CIL-1, $[\alpha]_{589}^{20} = -44.3°$

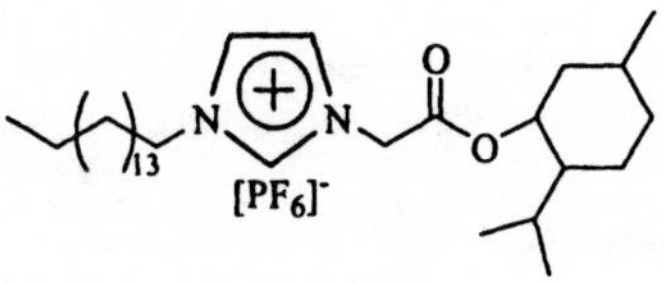

CIL-2, $[\alpha]_{589}^{20} = -31.0°$

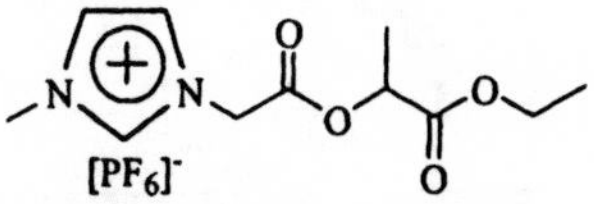

CIL-3, $[\alpha]_{589}^{20} = -30.3°$

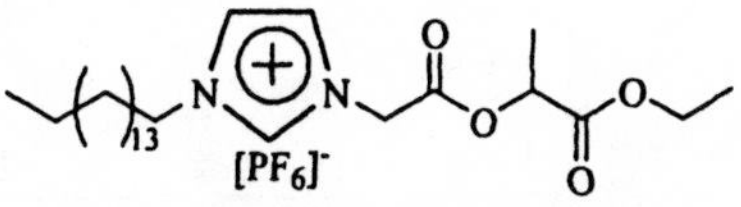

CIL-4, $[\alpha]_{589}^{20} = -17.8°$

The four chiral ionic liquids were synthesized via a similar synthetic route, which is depicted in Scheme 3 by taking the preparation of CIL-1 as an example. The specific rotations of all four chiral ionic liquids were measured in tetrahydrofuran with a concertration of 0.02 g/ml. The chemical structures were proved by 400 ^{1}H NMR.

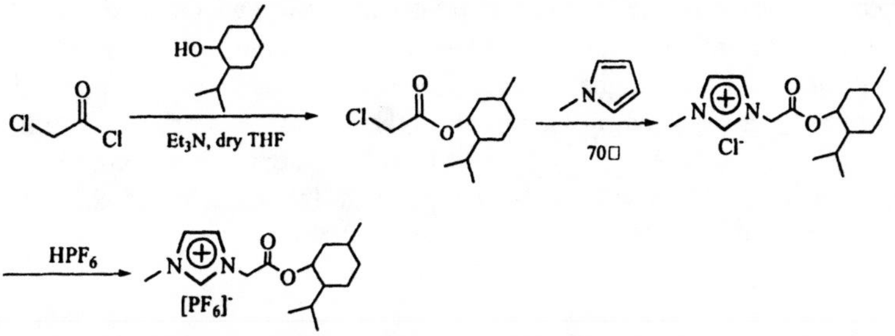

Scheme 3. Synthetic route of chiral ionic liquids CIL-1

Reverse ATRP of MMA was carried out in CIL-1 and CIL-2 at 80 □. The polymerization results are summarized in Table 5. During the same reaction time, the monomer conversion in bulk was comparable to those in chiral ionic liquids. However, the molecular weight distribution of the obtained polymer in the absence of ionic liquids was much wider than those of PMMAs obtained in IIs. An ill-controlled reaction process is indicated. It is consistent with the observation by Matyjaszewski et al. that AIBN/CuCl$_2$/bipy initiating system can't promote a controlled radical polymerization of MMA in bulk. The good solubility of catalyst in both CIL-1 and CIL-2 again help to mediate the polymerization process.

Table 5 Reverse ATRP of MMA in chiral ionic liquids[a]

Entry	Solvent	M$_n$×10^{-4} (Dalton)	M$_w$/M$_n$	Yield (%)
1	none	2.43	2.07	78.6
2	CIL-1	2.24	1.16	73.3
3	CIL-2	1.67	1.18	72.0

[a] [MMA]:[AIBN]:[CuCl$_2$]:[bipy] = 100:1:2.4:10, [MMA]/[CIL] = 3/1, T = 80□, time = 8h.

Source: Reproduced with permission from reference 53. Copyright 2003 Springer.

The asymmetric environments of chiral ionic liquids were found to exert some imfluence on the stereostructure of the PMMA obtained via radical polymerization. Table 6 showed the polymerization results in CIL-3. PMMA obtained by reverse ATRP has a narrow polydispersity, which is consistent with the results in the aforementioned reverse ATRP systems. In addition, all the polymers obtained have higher isotacticities than those in non-chiral media. A similar result was reported by Kubisa recently.[54] In their work, the chiral ionic liquid, 1-(R-(+)-2'-mehtylbutyl)-3-methylimidazolium hexafluorophosphate ([mbmim*][PF$_6$]), was used as the media, and the tacticity of the obtained PMMA with a high isotacticity measured by ^{13}C NMR.

Table 6 Reverse ATRP of MMA in chiral ionic liquid CIL-3[a]

Entry	M$_{n,GPC}$ ×10^{-4}, Dalton	M$_{n,NMR}$ ×10^{-4}, Dalton	PDI	Tacticity			Yield, %
				mm	mr	rr	
1	1.54	2.31	1.21	8.32	37.22	54.46	75.65
2[b]	1.62	2.09	1.39	8.12	37.91	53.97	64.00
3[c]	6.41	-	2.76	7.17	38.45	54.38	74.42

[a] [MMA]:[AIBN]:[CuCl$_2$]:[bipy] = 100:1:2:4, [MMA]/[CIL] = 3:1, T = 90°C, time = 24h.

[b] In the recovered CIL-3 and catalytic system.

[c] Conventional radical polymerization in CIL-3.

Conclusions

In summary, ionic liquids are good alternative solvents in reverse ATRP of MMA. The initiating system, AIBN/CuCl$_2$/bipy, which had been reported as unable to promote a controlled process of MMA in bulk and solution, mediates the radical polymerization of MMA to proceed in a controlled way in all ionic liquids used. The other advantages of ionic liquids in comparison with commonly used organic solvents include the reduced amount of catalyst needed, easy separation of resultant polymers, contamination prevention of the resultant polymer, and recyclability of the catalyst system. Moreover, it is possible to control the tacticity of the polymer obtained via radical polymerization by a combination of carefully chosen reaction condition and well-designed ionic liquids. With the rapid development of green chemistry and the expansion of ionic liquids family, ionic liquids will play more and more important roles in chemical research and industry. It is reasonable to predict that the novel type solvents will work better for us in the future when various general and functionized ionic liquids are prepared.

Acknowledgments

This work is supported by the National Natural Science Foundation of China (Grants 20174001, 20325415) and the Chinese Ministry of Education through the teaching and research award fund for outstanding young teachers in higher education institutions.

References

1. Seddon KR. In "Ionic Liquids in Synthesis". Wasserscheid P, Welton T Eds. Wiley-VCH: Weinheim. Preface, 2003.
2. Janz GJ. "Molten Salts Handbook". Academic Press: New York. Preface, v, 1967.
3. Welton T. "Room temperature ionic liquids. Solvents for synthesis and catalysis". Chemical Reviews, 1999, 99(8): 2071-2083.
4. Seddon KR. "Ionic liquids for clean technology". Journal of Chemical Technologe and Biotechnologe, 1997, 68(4): 351-356.
5. Matsumoto A. In "Handbook of Radical Polymerization". Matyjaszewski K. and Davis TP Eds. John Wiley & Sons: New York. Chapter 13, 2002.
6. Georges MK, Veregin RPN, Kazmaier PM, Hamer GK. "Narrow molecular-weight resins by a free-radical polymerization process". Macromolecules, 1993, 26(11): 2987-2988.

7. Hawker CJ. "Molecular-weight control by a living free-radical polymerization process". J. Am. Chem. Soc., 1994, 116(24), 11185-11186.

8. Wang JS, Matyjaszewski K. "Controlled living radical polymerization promoted by a Cu(I)/Cu(II) redox process". Macromolecules, 1995, 28(23): 7901-7910.

9. Wang JS, Matyjaszewski K. "Controlled living radical polymerization - atom-transfer radical polymerization in the presence of transition-metal complexes". Journal of the American Chemistry Society, 1995, 117(20): 5614-5615.

10. Patten TE, Xia JH, Abernathy J, Matyjaszewski K. "Polymers with very low polydispersities from atom transfer radical polymerization". Science, 1996, 272: 866-868.

11. Wang JS, Matyjaszewski K. "Living Controlled Radical Polymerization-Transition-Metal-Catalyzed Atom-Transfer Radical Polymerization in the Presence of a Conventional Radical Initiator". Macromolecules, 1995, 28(22): 7572-7573.

12. Xia JH, Matyjaszewski K. "Controlled/"living" radical polymerization. Homogeneous reverse atom transfer radical polymerization using AIBN as the initiator". Macromolecules, 1997, 30(25): 7692-7696.

13. Kato M, Kamigaito M, Sawamoto M, Higashimurat. "Polymerization of methyl methacrylate with the carbon tetrachloride dichlorotris(triphenylphosphine)ruthenium (II) methylaluminum bis(2,6-di-tert-butylphenoxide) initiating system possibility of living radical polymerization". Macromolecules, 1995, 28(5): 1721-1723.

14. Krstina J, Moad G, Rizzardo E, Winzor CL, Berge CT, Fryd M. "Narrow polydispersity block-copolymer by free-radical polymerization in the presence of macromonomers". Macromolecules, 1995, 28: 5381-5385.

15. Krstina J, Moad G, Rizzardo E, Winzor CL, Berge CT, Fryd M. "A new form of controlled growth free radical polymerization". Macromol. Symp., 1996, 111: 13-23.

16. Moad MG, Moad CL, Krstina J, Rizzardo E, Berge CT, Darling TR. WO 96/15157 (1996).

17. Chen XP, Qiu KY. "Synthesis of well-defined poly(methyl methacrylate) by radical polymerization with a new initiation system TPED/FeCl$_3$/PPh$_3$". Macromolecules, 1999, 32(26): 8711-8715.

18. Qin DQ, Qin SH, Qiu KY. "A reverse ATRP process with a hexasubstituted ethane thermal iniferter diethyl 2,3-dicyano-2,3-di(p-tolyl)succinate as the initiator". Macromolecules, 2000, 33(19): 6987-6992.

19. Haddleton DM, Jasieczek CB, Hannon MJ, Shooter AJ. "Atom transfer radical polymerization of methyl methacrylate initiated by alkylbromide and 2-pyridinlcarbaldehyde imine". Macromolecules, 1997, 30(17): 2190-2193.

20. Haddleton DM, Crossman MC, Dana BH, Cuncalf DJ, Heming AM, Kukulj D, Shooter AJ. "Atom transfer polymerization of methyl methacrylate mediated by alkylpyridylmethanimine tape ligands, copper(I) bromide, and alkyl halides in hydrocarbon solution". Macromolecules, 1999, 32(7): 2110-2119.

21. Pino P, Suter UW. "Some aspects of stereoregulation in the stereospecific polymerization of vinyl monomers". Polymer, 1976, 17(11): 977-995.

22. Nakano T, Okamoto Y. "Asymmetric polymerization of methacrylates". Macromol. Rapid Commun., 2000, 21(10): 603-612.

23. Isobe Y, Nakano T, Okamoto Y. "Stereocontrol during the free-radical polymerization of methacrylates with Lewis acids". J. Polym. Sci. Polym. Chem., 2001, 39(9): 1463-1471.

24. Isobe Y, Yamada K, Nakano T, Okamoto Y. "Stereospecific free-radical polymerization of methacrylates using fluoroalcohols as solvents". Macromolecules, 1999, 32: 5979-5981.

25. Huddleston JG, Willauer HD, Swatloski RP, Visser AE, Rogers RD. "Room temperature ionic liquids as novel media for 'clean' liquid-liquid extraction". Chemical Communications, 1998, 16: 1765-1766.

26. Hong HL, Zhang HW, Mays JW, Visser AE, Brazel CS, Holbrey JD, Reichert WM, Rogers RD. "Conventional free radical polymerization in room temperature ionic liquids: a green approach to commodity polymers with practical advantages". Chemical Communications, 2002, 13: 1368-1369.

27. Harrisson S, Mackenzie SR, Haddleton DM. "Unprecedented solvent-induced acceleration of free-radical propagation of methyl methacrylate in ionic liquids". Chemical Communications, 2002, 23: 2850-2851.

28. Ryan J, Aldabbagh F, Zwtterlund PB, Yamada B. "First nitroxide-mediated controlled/living free radical polymerization in an ionic liquid". Macromol. Rapid Commun., 2004, 25(9): 930-934.

29. Bottino FA, Fabbri E, Fragala IL, Malandrino G, Orestano A, Pilati F, Pollicino A. "Polystyrene-clay nanocomposites prepared with polymerizable imidazolium surfactants". Macomol. Rapid Commun., 2003, 24(18): 1079-1084.

30. Snedden P, Cooper AI, Scott K, Winterton N. "Cross-linked polymer-ionic composite materials". Macromolecules, 2003, 36(12): 4549-4556.

31. Perrier S, Davis TP, Carmichael AJ, Haddleton DM. "First report of reversible addition-fragmentation chain transfer (RAFT) polymerisation in room temperature ionic liquids". Chemical Communications, 2002, 19: 2226-2227.

32. Carmichael AJ, Haddleton DM, Bon SAF, Seddon KR. "Copper(I) mediated living radical polymerisation in an ionic liquid". Chemical Communications, 2000, 14: 1237-1238.

33. Sarbu T, Matyjaszewski K. "ATRP of methyl methacrylate in the presence of ionic liquids with ferrous and cuprous anions". Macromolecular Chemistry and Physics, 2001, 202(17): 3379-3391.

34. Biedron T, Kubisa P. "Atom-transfer radical polymerization of acrylates in an ionic liquid". Macromol. Rapid Commun., 2001, 22(15): 1237-1242.

35. Biedron T, Kubisa P. "Atom transfer radical polymerization of acrylates in an ionic liquid: Synthesis of block copolymers". J. Polym. Sci. Polym. Chem., 2002, 40(16): 2799-2809.

36. Zhao YL, Chen CF, Xi F. "Living/controlled radical copolymerization of N-substituted maleimides with styrene in 1-butyl-3-methylimidazolium hexafluorophosphate and anisole". J. Polym. Sci. Polym. Chem., 2003, 41 (14): 2156-2165.

37. Zhao YL, Zhang JM, Jiang J, Chen CF, Xi F. "Atom transfer radical copolymerization of N-hexylmaleimide and styrene in an ionic liquid". J. Polym. Sci. Polym. Chem., 2002, 40(20): 3360-3366.

38. Ma HY, Wan XH, Chen XF, Zhou QF. "Reverse atom transfer radical polymerization of methyl methacrylate in room-temperature ionic liquids" J. Polym. Sci. Polym. Chem., 2003, 41(1): 143-151.

39. Dupont J, de Souza RF, Suarez PAZ. "Ionic liquid (molten salt) phase organometallic catalysis". Chemical Reviews, 2002, 102(10): 3667-3691.

40. Branco LC, Rosa JN, Ramos JJM, Afonso CAM. "Preparation and characterization of new room temperature ionic liquids". Chemistry of A European Journal, 2002, 8(16): 3671-3677.

41. Ma HY, Wan XH, Chen XF, Zhou QF. "Reverse atom transfer radical polymerization of methyl methacrylate in imidazolium ionic liquids". Polymer, 2003, 44(18): 5311-5316.

42. Ma HY, Wan XH, Zhou QF. "The Study on Free Radical Polymerization of Methyl Methacrylate in Ionic Liquids". Ph. D. thesis, Library of Peking University, 2003, p30-33.

43. Freemantle M. "Designer solvents - Ionic liquids may boost clean technology development". Chemical & Engineering News, 1998, 76(13): 32-37.

44. Morrison JD, Mosher HS. In "Assymetric Organic Reactions". Prentice-Hall Inc: Englewood Cliffs, New Jersey, Chapter 10, 1971.

45. Howarth J, Hanlon K, Fayne D, McCormac P. "Moisture stable dialkylimidazolium salts as heterogeneous and homogeneous Lewis acids in the Diels-Alder reaction". Tetrahedron Letter, 1997, 38(17): 3097-3100.

46. Earle MJ, McCormac PB, Seddon KR. "Diels-Alder reactions in ionic liquids - A safe recyclable alternative to lithium perchlorate-diethyl ether mixtures". Green Chemistry, 1999, 1(1): 23-25.

47. Wasserscheid P, Bösmann A, Bolm C. "Synthesis and properties of ionic liquids derived from the 'chiral pool'". Chemical Communications, 2002, 3: 200-201.
48. Levillain J, Dubant G, Abrunhosa I, Gulea M, Gaumont AC. "Synthesis and properties of thiazoline based ionic liquids derived from the chiral pool". Chemical Communications, 2003, 23: 2914-2915.
49. Ishida Y, Miyauchi H, Saigo K. "Design and synthesis of a novel imidazolium-based ionic liquid with planar chirality". Chemical Communications, 2002, 19: 2240-2241.
50. Bao WL, Wang ZM, Li YX. "Synthesis of chiral ionic liquids from natural amino acids". Journal of Organic Chemistry, 2003, 68(2): 591-593.
51. Ujiie S, Iimura K. "Ion complex type of novel chiral smectic C* liquid crystal having chiral hydrogentartrate counterion". Chemistry Letters, 1994, 17-20.
52. Haramoto Y, Miyashita T, Nanasawa M, Aoki Y, Nohira H. "Liquid crystal properties of new ionic liquid crystal compounds having a 1,3-dioxane ring". Liquid Crystal, 2002, 29(1): 87-90.
53. Ma HY, Wan XH, Chen XF, Zhou QF. "Design and synthesis of novel chiral ionic liquids and its application in free radical polymerization of methyl methacrylate" Chinese Journal of Polymer Science, 2003, 21(3): 265-270.
54. Biedron T, Kubisa P. "Ionic liquids as reaction media for polymerization processes: atom transfer radical polymerization (ATRP) of acrylates in ionic liquids". Polymer International, 2003, 52(10): 1584-1588.

Chapter 5

Applying Ionic Liquids for Controlled Processing of Polymer Materials

John D. Holbrey, Ji Chen, Megan B. Turner, Richard P. Swatloski, Scott K. Spear, and Robin D. Rogers

Center for Green Manufacturing and Department of Chemistry, The University of Alabama, Tuscaloosa, AL 35487

This perspective examines the potential, highlighting some examples from our on-going research program, to evaluate and apply ionic liquids as advanced functional solvents for dissolving and processing polymers in order to prepare active materials and composites for sensor and smart materials applications integrating complexants or colorimetric and biological receptors into biorenewable, biocompatable substrate matrices.

Introduction

Ionic Liquids (ILs), composed solely of ionized species, have applications in a range of electrochemical, synthetic, catalytic, and separations technologies (*1,2*). Much of the interest in ILs derives from their property sets (especially lack of vapor pressure) which lends them to green chemistry approaches, reducing the reliance on volatile organic compounds (VOCs) as solvents. ILs have proven to be useful solvents for a variety of polymer applications; for radical, ionic and electrochemical polymerizations, *in situ* formation of polymer-electrolyte membranes, and as solvents for dissolution and processing of polymers such as cellulose which is generally regarded as *difficult* to dissolve and process. However, while it has been reported that ILs are exceptional solvents for almost all materials (*3*), it is important to make the distinction between the properties of any one specific example of an IL and the generic behavior of the class of solvents. Not all materials will dissolve in all ILs, and not all ILs display the same properties or characteristics.

Early investigations were initially limited to using ILs as acid catalysts for oligomerization and polymerization of olefins (*4*) and for synthesis of polyanilines (*5*) and polyphenylenes (*6*). Transition metal catalysts were also demonstrated in these acidic systems, for example, in Ziegler-Natta catalysis (*7*) and polyethylene formation using Ni-diimine catalysts (*8*). Subsequent development of air and moisture stable ILs led to increased utility of ILs for polymer applications, including radical polymerization reactions (*9*).

Polymer supported electrolytes are becoming increasingly important for modern battery, fuel cell, and photoelectric devices. Especially for lithium-metal-polymer batteries to power consumer devices. Watanabe (*10*) used free-radical polymerization of 2-hydroxymethyl-methacrylate to prepare IL-in-polymer conducting electrolyte films. Carlin and Fuller (*11*) synthesized catalytic membranes for heterogeneous hydrogenation incorporating palladium in gas-permeable IL–polymer gel membranes. The IL/polymer membranes were flexible and dimensionally stable with the IL providing ionic conductivity and flexibility to the otherwise insulating, rigid PVdF(HFP) copolymer. It has been suggested that application of ILs leading to higher conductivity, lack of evaporation, and plasticizing effects, will be the next significant advance in practical polymer electrolyte technologies (*12*). The ability of ILs to act as plasticizers for polymer matrixes has been more recently investigated in a systematic way by Brazel and co-workers, exploring the use of ILs as plasticizers for poly(methyl methacrylate) (*13*) leading to insight into the dual application of ILs in these systems.

ILs can also be used as structure-creating solvents for preparing polymeric materials. Winterton and co-workers (*14*) have shown how polymer composites, some with permanent porosity, can be prepared by monomer *in-situ* polymerization, using ILs as the solvents and presumably also as porogenic templating agents. Polymer-IL gels have been synthesized by cross linking reactions of two functionalized poly(ethylene glycol) oligomers in a hydrophobic IL (*15*), see Figure 1.

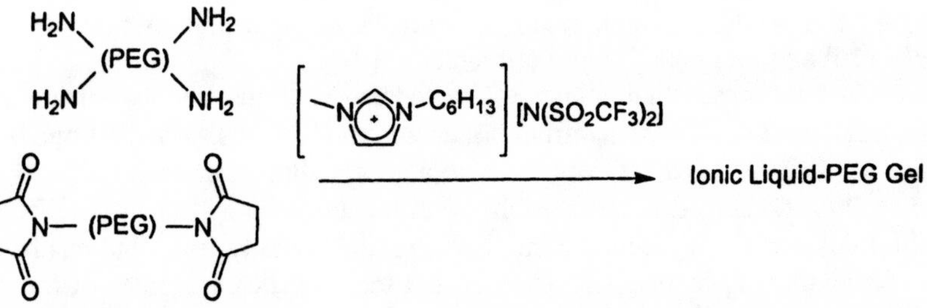

*Figure 1. Formation of structured IL-PEG gels in the hydrophobic
IL, 1-hexyl-3-methylimidazolium bis(trifluoromethanesulfonyl)imide (15).*

Cellulose Solubility in Ionic Liquids

Cellulose (Figure 2) h as m any i mportant commercial applications in the fiber,
paper, membrane, polymer, and paints industries. A restriction on the use of
underivatized cellulose, and a cause of major environmental pollution, is the
limited number of common solvents in which cellulose is soluble; currently,
cellulose processing and chemistry relies largely on the use of carbon disulfide
(CS_2) and caustic bases as dissolving solutions to prepare soluble cellulose
xanthate derivatives as intermediates. The efficiency of existing methods for
dissolving and derivatizing cellulose can be significantly improved by the
availability of suitable solvents for refined and natural cellulose; such an
example is N-methylmorpholine-N-oxide (NMMO), used as a solvent for
non-derivatizing dissolution of cellulose for the production of lyocell fibers.

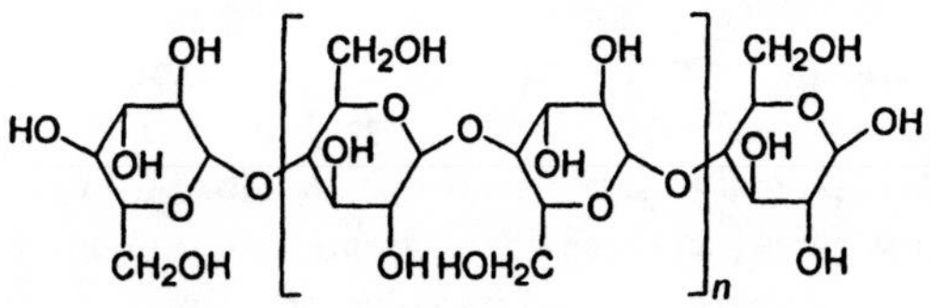

Figure 2. Molecular structure of cellulose.

We have demonstrated that some ionic liquids, in particular,
1-butyl-3-methylimidazolium chloride ([C_4mim]Cl, Figure 3), a polymorphic
crystalline solid at room temperature (*16*), can be used in the molten state as a
non-derivatizing solvent to dissolve cellulose (*17-19*). In a presumably similar
manner, cellulose has also been shown to be dissolvable in a few examples of

higher temperature molten salts, including *N*-alkylpyridinium chloride molten salts (*20*) and inorganic lithium salt hydrates (*21*).

Most recently, Wu et al. have followed up our initial studies, confirming the work and also describing a variant imidazolium IL, 1-allyl-3-methylimidazolium chloride (Figure 3, *right*), which is also a good cellulose solvent (*22*). This can be easily rationalized in terms of the need for the good hydrogen-bond acceptor (chloride) and also hydrogen-bond donating functionality (contributions from the imidazolium ring hydrogens) although the mechanistics and rôles of the cation and anion are not yet clear. The authors have also recently reported the acetylation of cellulose in this IL system (*23*), analogous to the established reaction procedure in *N*-alkylpyridinium molten salts (*24*). The resultant cellulose derivatives having correspondingly high solubility in organic solvents chloroform and acetone.

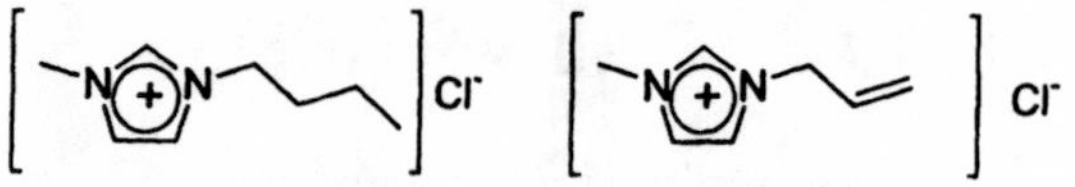

Figure 3. Structure of 1-butyl-3-methylimidazolium chloride (left) and 1-allyl-3-methylimidazolium chloride (right).

These current advances using lower melting imidazolium ILs appear to be very promising (*17-19,22,23*). The cellulose/IL solutions have been shown to be amenable to conventional processing techniques using the same approach taken by researchers investigating other cellulose solvents such as NMMO, thus making the formation of threads, thin films, and beads a relatively simple process. Figure 4 shows the SEM of a cellulose fiber prepared by pulling from a viscous solution containing ca. 10 wt% cellulose in molten [C$_4$mim]Cl into water.

In a significant study of the dissolution of cellulose by LiCl/dimethylacetamide solvent, Spange et al. (*25,26*) highlighted the polarity parameters, alpha, beta, and pi star, of various cellulose forms. Moderate values of π^* and beta were found whereas the hydrogen bond donor values were more significant. The study also examined by solvatochromatic methods the polarity parameters associated with typical solvents associated with cellulose dissolution. It was found that significant increases in hydrogen bond acidity values were found for the solvent dimethylacetamide when LiCl was added at high concentration (15%). The cellulose solvent and cellulose appear to have rather complimentary hydrogen bonding capabilities allowing the heat of mixing of the polymer and the solvent to compensate for the disruption of the crystal lattice (*26*).

*Figure 4. Scanning electron micrograph of a cellulose fiber pulled from IL
solution (prepared in collaboration with Prof. R. M. Broughton, Auburn
University, Auburn, AL).*

This interpretation is significantly strengthened by linear solvent-energy
relationship (LSER) results *(27,28)*, using Abraham's generalized solvent
equation *(29)* as shown in eq. 1 to model solvent characteristics, in which the *a*
coefficient, the solvent hydrogen bond basicity, is insignificant for
dimethylacetamide without LiCl in which cellulose is not soluble but reaches a
quite extreme value in [C$_4$mim]Cl in which cellulose is highly soluble (Table 1).

$$log\ K = c + rR_2 + s\pi_2^H + a\Sigma\alpha_2^H + b\Sigma\beta_2^H + v\mathrm{LogV} \qquad (1)$$

In eq. 1, R is the excess molecular refraction, V is the molecular volume, *a*
is the hydrogen bond acidity parameter, *b* is the hydrogen bond basicity
parameter, and *s* is the polarity/polarizability parameter. The LSER approach to
characterizing solvent properties is particularly useful for ionic liquid systems,
when compared to purely empirical methods such as through miscibility *(30)* or
the use of solvatochromic probes *(31)*, in that solvent characteristics are broken
down in terms of hydrogen bond donor and acceptor ability, and polarizability
characteristics, since any IL will contain at least two different ionic components
which can interact independently with both dissolved solutes *and* with the other
ions in the IL.

It is instructive to compare the values of the coefficients for selected ILs and
solvents (which have been used to dissolve cellulose) as shown in Table 1. It is
interesting to note that none of these solvents has particularly high values of the
v coefficient in comparison to systems where the distribution of solutes takes
place relative to water. Water is strongly ordered and its cohesivity much higher

than most solvents, but this is not the case when comparing these solvents to hexane. In general they all show greater order than hexane but the difference is not large.

The organic solvents show interaction through the π-system, however, for the ionic liquids this is quite variable and dependent on the anion with the chloride showing the greatest preference. The ionic liquids show no tendency to interact with hydrogen bonding bases, hydrogen bond acceptors, whereas the organic solvents do show weak interaction with these groups. However, most striking of all is the apparent strength of the interaction between the imidazolium chloride and hydrogen bond donors. This interaction is much weaker for the non-chloride ionic liquids and is completely absent for the other solvents. This distinction is particularly striking for dimethylacetamide which is an effective cellulose solvent, and its absence suggests that this functionality is supplied by the additional chloride added to this solvent.

Table 1. Some Typical Interaction Parameters of the Solvents at 100 °C, IL data from Ref. 27.

IL	r	s	a	b	v
[C$_4$mim]Cl	0.408	1.826	4.860	-0.121	0.392
[C$_4$mim][BF$_4$]	-0.141	1.365	1.660	-0.283	0.473
[C$_4$mim][PF$_6$]	0	1.540	1.369	0	0.439
Dimethylacetamide	0.36	1.33	0	0.78	0.79
Dimethylformamide	0.37	1.31	0	0.74	0.65
Dimethylsulfoxide	0.52	1.74	0	0.88	0.78

It is apparent that the LSER approach, particularly where the electronic interactions are highlighted against the van der Waals solvent hexane, is useful in determining the most significant of these. Complementarity of solute–solvent interactions is highly significant in predicting solubility of polymers in novel solvent systems and represents the current best approach to designing new, more efficient, and more environmentally benign approaches to polymer processing. In turn, knowledge of the solvent properties may be applied to the design of solvent systems tailored to the dissolution of selected solutes or polymers.

The ability to readily dissolve cellulose, and potentially also other 'difficult' polymer substrates, presents exciting new opportunities to enable the formation of functional composite materials. Here, we introduce our efforts to develop strategies using IL dissolution of cellulose as an enabling step to prepare functional composite materials for device applications by incorporating or introducing useful complexants, bioactive enzymes, and macroscopic additives to the cellulose to either change (improve) the properties of the materials, or to introduce some selective functionality.

Approaches to Cellulose Composite and Encapsulated Materials

As described in the previous section, cellulose can be readily dissolved without derivatization in [C₄mim]Cl by heating, especially using microwave heating to permit rapid dissolution. If, after dissolution, secondary components are added to the solution before regeneration, blended or composite materials can be prepared.

The simplicity and flexibility of this approach can be illustrated by demonstrating how dyes or complexants for coordination and binding of metal ions can be incorporated by dissolving the additives in IL solution and mixing with the cellulose solutions. Many dyes, and complexants that have been designed to be water-insoluble can be readily dissolved in the polar [C₄mim]Cl IL at high concentration and so integrated into a processed *hydrophilic* cellulose matrix to obtain materials suitable for sensing and remediation in aqueous media. Cellulose membranes may be advantageous over other synthetic polymers for these uses, with respect to their high porosity for water, and low (in the dry state) gas porosities. This hydrophilicity, or wetability can be very useful, for example in p roviding f ast t ransport o f w ater-soluble i ons (metals) to the active sensing sites. Because the IL is able to dissolve many water-insoluble materials, this also provides a methodology to entrap or incorporate them into the cellulose matrix in a highly dispersed manner.

Similarly, insoluble macromolecular particles including enzymes, and inorganic nanoparticle components can be introduced by dispersion, in the cellulose-in-IL solutions providing simple regeneration routes to incorporate structurally m odified c ellulose m aterials w ith possible uses in biocatalysis and sensing, magneto-responsive materials, and with additives to retard thermal and radiative degradation (for example for flame retardants and UV-filters, respectively). A number of dispersed particle composites are of major interest for smart, or enhanced materials.

There are two primary compositing techniques available for use: (i) dissolution of cellulose and one (or more) additional components allowing intimate blending and (ii) dispersion of additives, for example nanoparticles, as a way of introducing insoluble particles throughout the matrix. In practice, the IL solvent is very effective for this, probably because the polar, electrolytic nature of the liquid reduces the tendency of nanoparticles to self aggregate.

This methodology is described below with reference to three distinct examples, formation of magnetite/cellulose composites, preparation of bioactive supported laccase films, and preparation of cellulose/polyacrylonitrile composites, which illustrate the methodology and scope of the process.

78

Magnetite/Cellulose Composites (32)

Magnetic cellulosic materials have been used for separations (*33*), catalysis (*34*), biological controlled drug delivery (*35*), and imaging (*36*) applications. The main procedures for preparation magnetic cellulose materials are 'wet grinding' (*37*), lumen-loading (*38*), and *in situ* precipitation of magnetite from solution (*39*).

Magnetite cellulose composites were prepared by taking a solution of microcrystalline cellulose in IL (between 2-10 wt% cellulose) and homogeneously dispersing known concentrations of magnetite powder (particle size < 5 μm, from Aldrich) in the solutions by mixing and vortexing. Composites containing between 10-50 wt% Fe_3O_4:cellulose were prepared. The composites were then regenerated from water, resulting in the formation of products with an apparent homogeneous dispersion of metal oxide particles. No residual magnetite powder was observed in the aqueous washings of the reconstituted threads, indicating that all of the powder had been incorporated.

Figure 5 (*left*) shows the powder X-ray diffraction patterns of composites containing 10 and 30 wt% magnetite (B and C) compared to patterns for neat magnetite (D) and regenerated cellulose (A), showing that the composites contain discrete, unchanged magnetite particles dispersed in the cellulose matrix. The magnetic hysteresis loop of a characteristic 30 wt% magnetite-in-cellulose composite, measured by alternating gradient magnetometry is shown in Figure 5 (*right*). All the samples were ferrimagnetic, comparable to that of the initial Fe_3O_4 powder, with the remanence ratio indicative of a small superparamagnetic contribution.

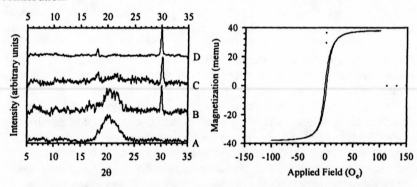

Figure 5. Magnetite/cellulose composites. X-ray powder diffraction spectra for (A) reconstituted microcrystalline cellulose; (B) 10 wt%, (C) 30 wt%, and (D) 100 wt% magnetite (left) and magnetic hysteresis loop for the 30 wt% magnetite-MCC material (right). Peak (220) was fit to a Gaussian curve with the data from 28-32 in 2θ.

Scanning electron microscopy (Figure 6) showed that the materials were homogenous in nature, with the magnetite particles dispersed evenly throughout the matrix. There appears to be a linear increase in magnetization with respect to magnetite loading, allowing one to make materials with a specific magnetic response. A ll t he m aterials p roduced e xhibited u niform l oading o f m agnetite, and with little evidence of expanded aggregation from the SEM and AGM results.

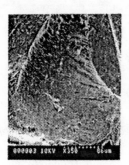  

Figure 6. Scanning electron micrographs of a 30 wt% magnetite/cellulose composite film, (left, ×350) showing film homogeneity, initial granularity of magnetite powder added (center, ×6000), and dispersed embedded magnetite particles in the cellulose film (right, ×7000).

Laccase Enzyme in Cellulose Polymer Films (40)

Cellulose appears to be an ideal support material for many enzyme systems, being both biosourced and largely biocompatible. Many examples of cellulose supported enzymes are known, in most cases, using cellulose-derivatives and covalent bonding via functional linkers to attach catalysts to the support surface. Entrapment, or encapsulation, without recourse to chemical attachment is certainly desirable if it can be achieved.

In order to determine whether the ready dissolution of cellulose in molten [C₄mim]Cl could be used as a processing route for encapsulation of enzymatic species in cellulose membranes or beads for reactive and sensing applications, two questions needed to be answered: (i) can encapsulated enzymes be prepared using the [C₄mim]Cl dissolution process, and (ii) are enzymes supported in such a cellulosic matrix still chemically active?

The first of these was addressed using the same approach as the addition of magnetite power, namely, enzyme particles can be freely dispersed in a cellulose in IL solution, and are retained within the super-structure of the regenerated matrix, b ased o n m aintenance o f c haracteristic UV signatures. Answering the second question of whether the entrapped enzymes retain biologic activity is of greater importance.

A model system, using laccase from *Rhus vernificera* (E.C. 1.10.3.2) was investigated (*9*). Laccase was chosen as a target enzyme because of its ready availability, and activity for aromatic oxidation in the presence of celluloses. Enzyme activity was determined by UV/Vis spectroscopy (*41*) following the laccase-catalyzed conversion of reduced syringaldazine (colorless, $\lambda_{max} = 371$ nm) to its oxidized form (pink, $\lambda_{max} = 555$ nm) from buffered aqueous solution (Figure 7).

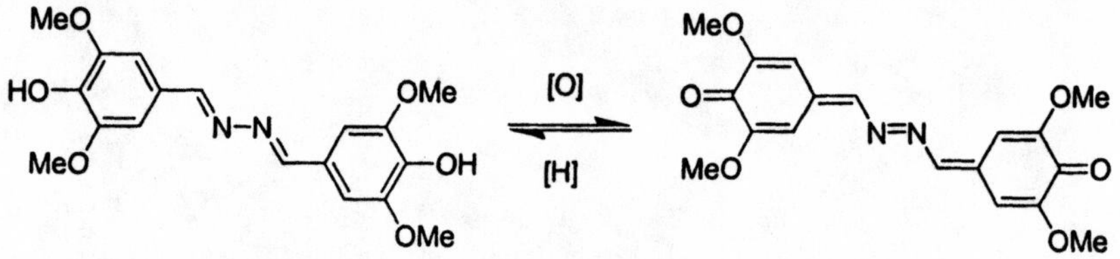

Figure 7. Reductive and oxidative conversion of syringaldazine between phenolic (colorless, left) and quinone (pink, right) forms.

Films initially prepared by dispersing laccase directly in the hot (~100 °C) cellulose/IL solution showed no activity. However, if the cellulose/IL solution is allowed to cool to room temperature, resulting in formation of a viscous super-cooled liquid, before dispersion of the enzyme, films could be prepared in which enzyme activity was observed in the assay procedure (Figure 8). This clearly shows that thermal shock is an important factor, and that the nature of the IL solution can be used advantageously to allow cold-processing and help retain some enzyme activity. However, the effects of IL-induced denaturation appeared to be the major factor in reducing activity (*42*).

The IL [C₄mim]Cl has been shown to have a significant denaturing effect on some enzymes, even in aqueous solution, due in part, to the intrinsically high Cl⁻ ion concentration and activity in the reaction media (*43*). One possible method for stabilizing enzymes in these systems might be to provide a 'protective' pre-coating of the enzyme. It has been suggested that coating enzymes in a hydrophobic IL can improve stability in organic media (*44*). To test whether IL-coating could help protect the enzyme in a second IL processing solution, laccase was introduced to the cellulose/[C₄mim]Cl solution after initially coating the laccase with the *hydrophobic* IL, 1-butyl-3-methylimidazolium bis(trifluoro-methanesulfonyl)imide. The system, containing the 'pre-coated' enzyme, not only retained activity for the syringaldazine oxidation reaction, but actually showed a ten-fold increase in concentration of product in the film.

These results show successful introduction of an active enzyme in cellulose film p repared u sing t he [C₄mim]Cl I L p rocessing r oute. Enzyme activity was demonstrated to be maintained by cold processing, a method made possible by the innate characteristics of the imidazolium ILs. Activity was improved by pretreatment of the enzyme, providing a less denaturing microenvironment, and was particularly enhanced with a hydrophobic IL coating.

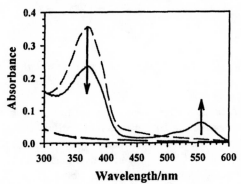

Figure 8. UV/Vis spectra showing laccase activity in IL-reconstituted cellulose films with reduction in the intensity of the peak at 371 nm corresponding to adsorbtion of the reduced form of syringaldazine and growth in the peak at 555 nm on production of the oxidized form.

Extension to Other Polymers and Formation of Polymer Composite Blends

We have suggested that the solubility of cellulose in [C₄mim]Cl is a consequence of chloride activity in the IL system which can disrupt inter-strand hydrogen-bonding of the cellulose polymers. Polarity measurments and solvatochromatic studies support this, and indicate that molten [C₄mim]Cl is a highly polar solvent, similar in solvent properties to DMF or DMSO. This suggests that other solutions of 'insoluble' or poorly soluble polymers, and also polymer blends, can be prepared, although attempting to correlate Hansen solubility parameters of polymers with solvent properties and Abraham solubility parameters of the ILs can be difficult, if not impossible, as discussed earlier.

Many common synthetic polymers (Figure 9) such as poly(vinyl alcohol) (PVA), poly(2-hydroxyethyl methacrylate) (PHEMA), polyacrylonitrile (PAN), polyaniline (emeraldine base) (PANI), polypropylene glycol (PPG), poly ethylene glycol (PEG), and biopolymers such as bovine serum albumin (BSA), deoxyribonucleic acid (DNA) salts, dextran, and starches have been found to be soluble in ILs such as [C₄mim]Cl, which incorporate strongly hydrogen bond accepting anions. However, solvation in ILs is not yet completely understood and some polymers with similar Hansen parameters can have markedly different solubility, for example PVC is soluble in [C₄mim]Cl whereas polystyrene is not.

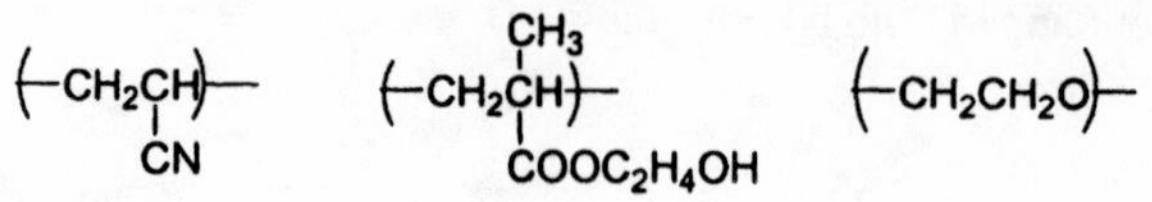

$$\left(\!-CH_2CH\!-\right)\qquad\left(\!-CH_2CH\!-\right)\qquad\left(\!-CH_2CH_2O\!-\right)$$
$$\underset{CN}{\big|}\qquad\qquad\underset{COOC_2H_4OH}{\overset{CH_3}{\big|}}$$

PAN   PHEMA   PEG

Figure 9. Structures of some polymers soluble in [C₄mim]Cl.

Cellulose/Polyacrylonitrile Blends

Cellulose/polymer blends have wide application, particularly in enhancing performance characteristics by introducing cellulose as the core blend component.   For example, cellulose/3-hydroxybutyrate-3-hydroxyvalerate copolymer (PHB/V) blend fibers show increased tensile strength from 20 MPa to 128-278 MPa compared with regular cellulose fibers.

One polymer of interest to us is the high temperature thermoplastic polyacrylonitrile, which is soluble only in a limited number of solvents including dimethylformamide, dimethylsulfoxide, and butyrolactone. The solubility of polyacrylonitrile in [C₄mim]Cl was investigated, and it was found to be readily soluble at 100 °C, forming clear viscous solutions up to 10 wt% polymer, assessed from the formation of optically clear solutions after dispersal of polymer and heating for 1 hour at 110 °C without stirring. Solutions containing between 5-10 wt% of polymer are viscous liquids, tending towards gels at the higher concentration range and can be drawn into fibers and regenerated from water, or other liquids miscible with the IL, yet non-solvents for the polymer, including acetonitrile and ethanol.

Composite   polymers   were   generated   by   mixing   solutions   of polyacrylonitrile and cellulose in IL, then extruding into water. The polymer was then washed with water, air dried and examined.   5 wt% cellulose (DP = 1056) and 2 wt% polyacrylonitrile (Mw = 86,000) solutions were prepared in [C₄mim]Cl. Dissolution was achieved with mixing at 104 °C over a 12 h time period. The two solutions were then mixed at 104 °C in varying proportions; yielding relative composition ranges of the two polymers from 20/80 to 80/20, as a ratio of weight percent of cellulose to polyacrylonitrile. Next the blended solutions were allowed to cool, and then coagulated as membranes using water. The resulting films were placed in a water bath and allowed to soak for 24 h in order to allow the maximum amount of IL to diffuse from the blended composite. Finally the composites were washed several times with water. The resulting soft, flexible cellulose/polyacrylonitrile membranes were dried in the oven for 24 h. As the water was evaporated, the films began to shrink t o f orm h ard, p orous m embranes. T he final films were analyzed using SEM (Figure 10), and DSC.

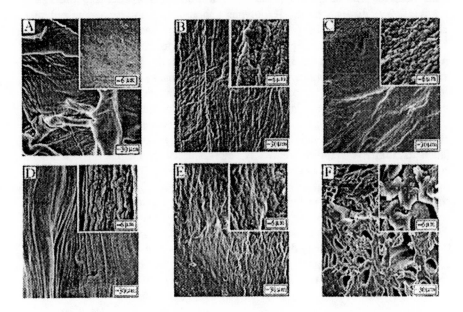

Figure 10. Scanning Electron Micrographs of cellulose (dissolving pulp, DP = 1046)/polyacrylonitrile blends: A: regenerated cellulose; B: cellulose/polyacrylonitrile; 20/80 (weight ratio); C: cellulose/polyacrylonitrile 40/60; D: cellulose/polyacrylonitrile 60/40; E: Cellulose/polyacrylonitrile 80/20; F: Regeneration polyacrylonitrile (×500 and ×5000, inset), showing the controlled changes in morphology.

On examination of the photographs, it appears that the surface is homogenous—indicating a miscible blend at all ratios from B to E. The blended materials all have different textures to that of the pure cellulose (A) or the pure PAN (F). The S EM p ictures s how t hat h omogeneous, m iscible s tructures a re formed across a range of blend compositions, and suppression of the characteristic porous structure of pure polyacrylonitrile, leading to denser, tougher composites even when as little as 20% cellulose was added.

Summary

We have demonstrated an approach to preparing cellulose supported materials with designed characteristics. We have introduced inorganic and biological particles by dispersion in IL/polymer solutions, and prepared polymer composites with modified properties. We have successfully shown that it is possible to produce homogenous magnetic composites by directly adding magnetite to cellulose containing IL solutions, and the formation of cellulose supported enzyme membranes for catalysis and sensing. Ongoing research is focussed on developing processing methods to prepare functional composite materials for structural, sensing and separations applications.

While ILs have some characteristics of relevance to a general 'green chemistry' approach, including a lack of volatility, almost all other properties (such as toxicity, stability, reactivity, etc.) vary with the cation and anion components present and can not be generalized. The utility of ILs, as a class of fluids, rests with individual examples displaying new, improved, or *different* combinations of solvent properties to other, existent liquids. It is important to remember that although ILs may appear to have similar solvent parameters, and thus seem to be *like* these molecular solvents, it is not the *same*, and this can lead to new, different, and hopefully, better chemistries and processes than other existing liquids. With regard to polymer composites, the unique and controllable solvent properties of ILs may be used to prepare materials from components that are mutually incompatible with other solvent systems.

Acknowledgments

This research is supported by the U.S. Environmental Protection Agency's STAR program through grant number R-83143201-0, U.S. Department of Energy, Division of Chemical Sciences, Geosciences, and Biosciences, Office of Basic Energy Research (Grant DE-FG02-96ER14673), and the PG Research Foundation.

References

1   *Ionic Liquids in Synthesis*; Wasserscheid, P., Welton, T., Eds.; VCH-Wiley: Weinheim, 2002.

2   Welton, T. *Chem. Rev.* **1999**, *99*, 2071; Holbrey, J. D.; Seddon, K. R. *Clean Prod. Proc.* **1999**, *1*, 223; Wasserscheid P.; Keim, W. *Angew. Chem. Int. Ed.* **2000**, *39*, 3772; Gordon, C. M. *Appl. Catal. A* **2001**, *222*, 101; Olivier

Bourbigou, H.; Magna, L. *J. Mol. Catal. A: Chem.* **2002**, *419*, 182; Dupont, J.; de Souza, R. F.; Suarez, P. A. Z. *Chem. Rev.* **2002**, *102*, 3667; Sheldon, R. *Chem. Commun.* **2001**, 2399; Dyson, P. J. *Appl. Organomet. Chem.* **2002**, *16*, 495; Song, C. E. *Chem. Commun.* **2004**, 1033.

3 Seddon, K. R. *J. Chem. Technol. Biotechnol.* **1997**, *68*, 351.

4 Abdul-Sada, A. K.; Atkins, M. P.; Ellis, B.; Hodgson, P. K. G.; Morgan, M. L. M.; Seddon, K. R. World Pat., WO95/21806, 1995; Hope, K. D.; Driver, M. S.; Harris, T. V. U.S. Pat 6,395,948, 2000.

5 Tang, J. S.; Osteryoung, R. A. *Synth. Met.* **1991**, *45*, 1; Tang, J. S.; Allendorfer, R. D.; Osteryoung, R. A. *J. Phys. Chem.* **1992**, *96*, 3531.

6 Kobryanskii, V. M.; Arnautov, S. A. *J. Chem. Soc., Chem. Commun.* **1992**, 727; Kobryanskii, V. M.; Arnautov, S. A. *Makromol. Chem.* **1992**, *193*, 455.

7 Carlin R. T.; Wilkes, J. S. *J. Mol. Catal.* **1990**, *63*, 125.

8 Pinheiro, M. F.; Mauler, R. S.; de Souza, R. F. *Macromol., Rapid Commun.* **2001**, *22*, 425

9 Kubisa. P. *Prog. Polym. Sci.* **2004**, *29*, 3.

10 Watanabe, M.; Mizumura, T. *Solid State Ionics* **1996**, *86-88*, 353.

11 Carlin, R. T.; Fuller, J. *Chem. Commun.* **1997**, 1345.

12 Shin, J.-H.; Henderson, W. A.; Passerini, S. *Electrochem. Commun.* **2003**, *5*, 1016.

13 Scott, M. P.; Brazel, C. S.; Benton, M. G.; Mays, J. W.; Holbrey J. D.; Rogers, R. D. *Chem. Commun.* **2002**, 1370; Scott, M. P.; Rahman, M.; Brazel, C. S. *Eur. Polymer J.* **2003**, *39*, 1947.

14 Snedden, P.; Cooper, A. I.; Scott, K.; Winterton, N. *Macromolecules* **2003**, *36*, 4549.

15 Klingshirn, M. A.; Spear, S. K.; Subramanian, R.; Holbrey, J. D.; Huddleston, J. G.; Rogers, R. D. *Chem. Mater.* **2004**, *16*, 3091.

16 Holbrey, J. D.; Reichert, W. M.; Nieuwenhuyzen, M.; Johnson, S. J.; Seddon, K. R.; Rogers, R. D. *Chem. Commun.* **2003**, 1636.

17 Swatloski, R. P.; Spear, S. K.; Holbrey J. D.; Rogers, R. D. *J. Am. Chem. Soc.* **2002**, *124*, 4974.

18 Swatloski, R. P.; Rogers, R. D.; Holbrey, J. D. '*Dissolution and processing of cellulose using ionic liquids*' U.S. Pat. App., 20030157351, August 21, 2003; Holbrey, J. D.; Spear, S. K.; Turner, M. B.; Swatloski, R. P.; Rogers, R. D. '*Cellulose matrix encapsulation and method*' U.S. Pat. App., 20040038031, February 26, 2004.

19 Swatloski, R. P.; Holbrey, J. D.; Spear S. K.; Rogers, R. D. "ILs for the dissolution and regeneration of cellulose." in *Molten Salts XIII, Proceedings of the Thirteenth International Symposium on Molten Salts,* Truelove, P. C.; De Long, H. C.; Mantz, R. A.; Stafford G. R.; Matsunaga, M; Eds.; 2003, 155.

20 Graenacher, C.; *Cellulose Solution*; U.S. Patent, 1943176, 1934.
21 Leipner, H.; Fischer, S.; Brendler, E.; Voigt. W. *Macromol. Chem. Phys.* **2000**, *201*, 2041.
22 Ren, Q.; Wu, J.; Zhang, J.; He, J. S.; Guo, M. L. *Acta Polymerica Sinica*, **2003**, 448.
23 Wu, J.; Zhang, J.; Zhang, H.; He, J.; Ren, Q.; Guo, M. *Biomacromol.* **2004**, *5*, 266.
24 Husemann, E.; Seifert, S. *Makromol. Chem.* **1969**, *128*, 288.
25 Spange, S.; Fischer, K.; Prause, S.; Heinze, T. *Cellulose* **2003**, *10*, 201.
26 Spange, S,; Reuter, A.; Vilsmeier, E.; Heinze, T.; Keutel, D.; Linert, W. *J. Polym. Sci., A* **1998**, *26*, 1945.
27 Anderson, J. L.; Ding, J.; Welton, T.; Armstrong, D. W. *J. Am. Chem. Soc.* **2002**, *124*, 14247.
28 Abraham, M. H.; Zissimos, A. M.; Huddleston, J. G.; Willauer, H. D.; Rogers, R. D.; Acree, W. E., Jr. *Ind. Eng. Chem. Res.* **2003**, *42*, 413.
29 Abraham, M. H. *Chem. Soc. Rev.* **1993**, *22*, 73.
30 Blanchard, L. A.; Brennecke, J. F. *Ind. Eng. Chem. Res.* **2001**, *40*, 287.
31 Carmichael, A. J.; Seddon, K. R. *J. Phys. Org. Chem.* **2000**, *13*, 591; Muldoon, M. J.; Gordon, C. M.; Dunkin, I. R. *J. Chem. Soc., Perkin Trans.* **2001**, 433.
32 Swatloski, R. P.; Holbrey, J. D.; Weston, J. L.; Rogers, R. D., *manuscript in preparation.*
33 Kurokawa, Y. *Desalination* **1982**, 41, 115; Al-Abdulla, I. H.; Mellor, G. W.; Childerstone, M. S.; Sidki, A. M.; Smith, D. S. *J. Immun. Methods* **1989**, *122*, 253.
34 Kurokawa, Y.; Hanaya, K. *Carbohydrate Polymers* **1995**, *27*, 313.
35 Rittich, B.; Španová, A.; Ohlashennyy, Y.; Lenfeld, J.; Rudolf, I.; Horák, D.; Beneš, M. J. *J. Chrom. B* **2002**, *774*, 25; Ito, R.; Machida, Y.; Sannan, T.; Nagai, T. *Int. J. Pharm.* **1990**, *61*, 109; Nagano, H.; Machida, Y.; Iwata, M.; Imada, T.; Noguchi, Y.; Matsumoto, A.; Nagai, T. *Int. J. Pharm.* **1997**, *147*, 119.
36 Matsumura, M.; Nakagami, H.; Chen, C. –J.; Ito, S.; Konno, T. *Pharma. Acta Helv.* **1998**, *73*, 205.
37 Kadwad, V. B.; Jyotsna, N.; Sivaprasad, N.; Sinha, P. K. *J. Radioanal. Nucl. Chem.* **1996**, *210*, 27.
38 Green, H. V.; Fox, T. J.; Scallan, A. M. *Pulp Pap. Can.* **1982**, *7*, 39.
39 Suber, L.; Foglia, S.; Ingo, G. M.; Boukos, N. *Appl. Organometal. Chem.* **2001**, *15*, 414.
40 Turner, M. B.; Spear, S. K.; Holbrey, J. D.; Rogers, R. D. *Biomacromol.* **2004**, *5*, 1379.

41 Harkin, J. M., Obst, J. R. *Science* **1973**, *180*, 296.

42 Hinckley, G.; Mozhaev, V. V.; Budde, C.; Khmelnitsky, Y. L. *Biotechnol. Lett.* **2002**, *24*, 2083.

43 Turner, M. B.; Spear, S. K.; Huddleston, J. G.; Holbrey, J. D.; Rogers, R. D. *Green Chem.* **2003**, *5*, 443.

44 Kim, M.-J.; Lee, J. K. *J. Org. Chem.* **2002**, *67*, 6845.

Chapter 6

Novel Solid Polymer Electrolytes Composed of Zwitterionic Liquids and Polymerized Ionic Liquids

Hiroyuki Ohno, Satoko Washiro, and Masahiro Yoshizawa

Department of Biotechnology, Tokyo University of Agriculture and Technology, Koganei, Tokyo 184–8588, Japan

Zwitterionic liquids (ZILs) are proposed to overcome the problem of component ion migration in ordinary ionic liquids (ILs). Both cation and anion are tethered to suppress their migration under potential gradient, and target ions can be transported by mixing ZILs and salts containing target ions. We have prepared novel solid polymer electrolytes composed of ZIL, LiTFSI, and polymerized IL, and investigated the effect on the ionic conductivity of adding ZIL. The ionic conductivity of the IL polymer fell with increasing LiTFSI concentration. According to Raman spectroscopy, the fraction of ion-pairs increased with increasing LiTFSI concentration. ZIL is confirmed as promoting the dissociation of LiTFSI in the polymer matrix.

Introduction

Ionic liquids (ILs) have many attractive properties including non-volatility, non-flammability, high ionic conductivity, wide potential window and thermal stability (1). They should therefore be useful in a variety of research areas. One application is in electrochemical devices such as rechargeable lithium-ion batteries (2-11), fuel cells (12-15), solar cells (16-21), and capacitors (22-27). However, where ILs as used as electrolyte solutions, they cannot transport the target ions alone, since component ions of the IL can also migrate along the potential gradient. We have proposed various methods to overcome this drawback and allow the transport of target ions alone. One strategy involves the preparation of zwitterionic liquids (ZILs) in which the cation and anion are tethered (28). After adding salt, only target ions are transported, since ZILs cannot migrate along the potential gradient. We have also designed polymerized ILs that realize target ion transport in an electrolyte film (29). However, it is still necessary to increase the dissociation of the added salt in order to generate more target ions. We therefore propose to include ZILs in polymerized ILs so as to improve the transport of target ions in polymerized ILs. Below, we investigate the effect of ion composition on the ion conducting behavior of some new solid polymer electrolytes consisting of polymerized ILs, LiTFSI, and ZILs.

Experimental

Materials. Tetra(ethylene glycol) diacrylate (>90%), N-ethylimidazole, 1,4-butanesultone, and 6-bromo-hexan-1-ol (>95%) were purchased from Tokyo Kasei Co. Acryloyl chloride (96%) and tri(ethylene glycol) divinyl ether (98%) were purchased from Aldrich Co. α,α'-Azobis(isobutyronitrile) (AIBN) was purchased from Kanto Chem. Co. AIBN was purified by recrystallization prior to use as an initiator for radical polymerization. LiTFSI (>99%) was a gift from Sumitomo 3M.

Synthesis of IL monomer and other components. Acryloyl chloride and an equimolar amount of 6-bromo-1-hexanol were mixed in tetrahydrofuran in the presence of triethylamine. The resulting bromohexyl acrylate was reacted with an excess of N-ethylimidazole. Imidazolium salt was prepared and was purified by the precipitation method with diethylether, and the bromide anion of the imidazolium salt was then replaced with the TFSI anion. We have previously reported the synthesis pathway of these materials in detail (29). We also reported the preparation of ZIL in our previous paper (28). Figure 1 shows the structure of IL monomer (**A**), cross-linkers (**B**), LiTFSI (**C**), and ZIL (**D**) .

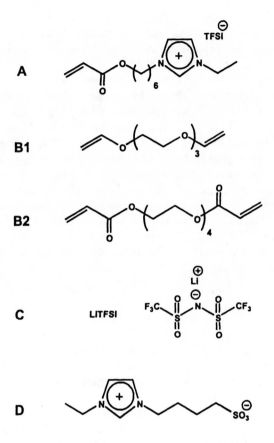

Figure 1. Structure of IL monomer, cross-linkers, LiTFSI, and ZIL.

Preparation of novel polymer gel electrolytes. IL monomer, AIBN, cross-linker, LiTFSI, and ZIL were mixed in the appropriate ratio in methanol so as to prepare a homogenous solution. The methanol was removed by evaporation and the mixture was poured then between two glass plates with the aid of a Teflon® spacer. The resulting material was polymerized in bulk under a N_2 atmosphere at 70 °C for 8 h. The polymers obtained were dried under a vacuum at 80 °C for 2 days.

Methods. The structure of the IL monomer and ZIL were confirmed by ^1H NMR spectroscopy (JEOL α-500). The ionic conductivity of the resulting polymers was measured by the complex-impedance method using an impedance analyzer (Solartron gain phase analyzer, model 1260, Schlumberger). The dynamic conductivity measurement system used was developed in our laboratory (30). All measurements were carried out in a glove box filled with dry N_2 in the temperature range 10 - 60 °C. Dynamic temperature changes were effected by a computer-controlled semi-conductor heat pump. The thermal behavior was measured using a DSC-6200 apparatus (SEIKO Instruments Inc.) in the temperature range -130 to +200 °C at a heating rate of 10 °C min^{-1}. To obtain the Raman spectra of the IL polymer containing LiTFSI and/or ZIL, samples were placed between glass plates. Raman spectra were recorded as a function of the LiTFSI concentration using a NRS-1000 Raman spectrometer (JASCO). The Kr$^+$ laser of wavelength 647.1 nm was irradiated with an Innova® 90C-K ion laser (COHERENT Inc.). As an interference filter, a holographic Super Notch® filter (KAISER Optical Systems Inc.) was used, and a SPEC-10™ System, liquid nitrogen-cooled CCD detector (Roper Scientific Inc.) was employed. The spectral resolution was 0.6 cm^{-1}. The Raman spectrometer was calibrated using indene and tetrachloromethane.

Results and discussion

We have already reported that IL polymers having flexible spacer between the imidazolium cation and vinyl group are effective in maintaining high ionic conductivity following polymerization of the corresponding IL monomers, such as A in Figure 1 (29). These are called IL-type polymer brushes. Although the polymer A had a relatively high ionic conductivity of about 10^{-4} S cm^{-1} at room temperature, the major carrier ions were TFSI anions in the polymer matrix. In this study, we added LiTFSI to the IL polymers to allow lithium cation transport in the IL polymers.

LiTFSI was added prior to polymerization of A in the presence of cross-linker B_1. Turbid solid polymer electrolytes were obtained irrespective of the LiTFSI concentration. These solid polymer electrolytes were obtained as films of up to 50 mol% LiTFSI content, but they became sticky and rubbery when the LiTFSI content exceeded 50 mol%.

To study the effect of adding LiTFSI on the ionic conductivity of the IL polymer, we measured the ionic conductivity of the corresponding mixtures. Figure 2 shows the temperature dependence of the ionic conductivity for polymerized A with B_1 as a cross-linker containing LiTFSI between 0 and 100 mol% relative to the imidazolium unit. The ionic conductivity of IL polymer containing no LiTFSI was approximately 10^{-4} S cm^{-1} at room temperature. The ionic conductivity changed little up to 10 mol%, but decreased monotonically with increasing LiTFSI concentration. Interestingly, Arrhenius plots of the ionic conductivity had almost the same slope whatever the LiTFSI concentration. In the case of polymer systems like polyether, the temperature dependence of the ionic conductivity become more pronounced with increasing salt concentration (31).

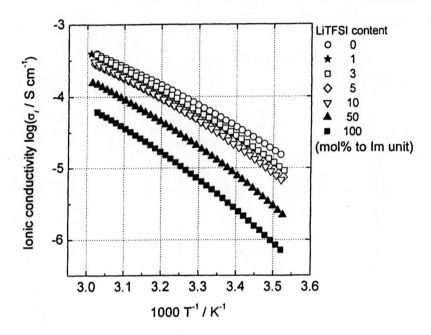

Figure 2. Temperature dependence of the ionic conductivity for polymerized A with B_1 containing various amounts of LiTFSI.

We investigated the effect of LiTFSI addition on the ionic conductivity and the glass transition temperature (Tg) of the IL polymer A with B_1. Figure 3 shows the effect of the LiTFSI concentration on the ionic conductivity at 30 °C, and Tg. The ionic conductivity decreases monotonically with increasing LiTFSI concentration. These findings suggest that the added LiTFSI does improve the ionic conductivity. In contrast, Tg behaved differently from the ionic conductivity, even though there is a strong relation between the ionic conductivity and Tg for ILs and their polymers (32). The value of Tg increased with increasing LiTFSI concentration up to 10 mol%, and then remained constant at about -50 °C over a wide range of LiTFSI concentrations. As stated above, the solid polymer electrolytes obtained were turbid systems beyond 50 mol% LiTFSI concentration. This suggests that Tg did not increase with increasing LiTFSI concentration at higher LiTFSI concentrations because the polymer system could not dissociate LiTFSI into ions at high concentrations. Transparent films were obtained at low LiTFSI concentrations, below 10 mol%. This may reflect the dissociation of LiTFSI in the polymer matrix, and is consistent with the increase in Tg at these concentrations.

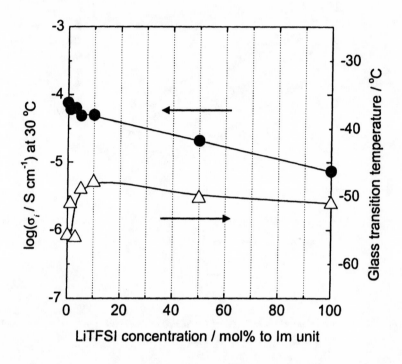

Figure 3. Effect of LiTFSI concentration on the ionic conductivity and Tg for IL polymer A with B₁.

Figure 4 shows Raman spectra of LiTFSI and IL polymer **A** with **B₁** as a function of the LiTFSI concentration over the frequency range 730 to 760 cm⁻¹. All Raman spectra were recorded at room temperature. The symmetric bending mode of the CF_3 group ($\delta_s CF_3$) for the TFSI anion is known to be at around 740 cm⁻¹ (33). The dissociation behavior of LiTFSI can be monitored by Raman spectroscopy from band shifts of $\delta_s CF_3$ in the region 740 – 750 cm⁻¹. The relative intensity and the peak position of $\delta_s CF_3$ depend on the LiTFSI concentration. For the sample with no LiTFSI, $\delta_s CF_3$ of IL polymer **A** was located at 739 cm⁻¹. The shape of this band became less symmetric and shifted to higher wavenumber with addition of salt to the IL polymer. A new band appeared at 746 cm^{-1} which agreed with the spectrum of solid LiTFSI, indicating that TFSI anions exist as ion pairs. However, the band at 739 cm^{-1} is attributed to free TFSI anions by Wang et al (34). Figure 4 shows that the fraction of ion pairs increases with increasing LiTFSI concentration.

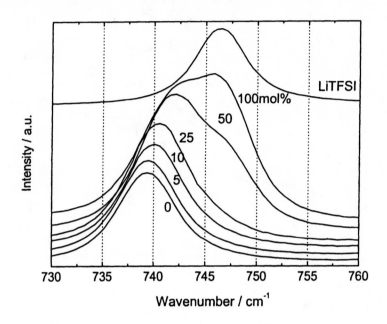

Figure 4. Raman spectra of IL polymer A with B$_1$ containing various amounts of LiTFSI.

From the data in Figure 4, the fraction of "free" and "ion paired" TFSI anions can be calculated. Figure 5 shows the dependence on concentration of the fraction of free and ion paired TFSI anions. No band splitting is observed at low concentrations, indicating that mostly there are free TFSI anions in the polymer matrix up to 5 mol% addition. After that, the fraction of free TFSI anions decreases monotonically decreased with increasing LiTFSI concentration. At LiTFSI concentrations exceeding 80 mol%, the fraction of ion paired TFSI anions is larger than that of free TFSI anions. Although the IL polymer is able to solubilize LiTFSI, the degree of dissociation is unexpectedly low in the polymer.

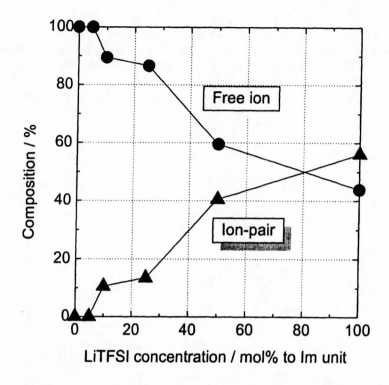

Figure 5. LiTFSI concentration dependence of the fraction of free and ion paired TFSI anions based on the fitting results of Raman bands.

Effect of ZIL addition

Figure 6 is a photograph of solid polymer electrolytes composed of polymerized **A**, **B₂**, **C** (LiTFSI: 50 mol% to Im unit), and differing amounts of ZIL (**D**). Without ZIL, the resulting polymer was a turbid sticky rubber as shown in the leftmost sample. Approximately 40% of LiTFSI salts were not dissociated when 50 mol% LiTFSI was added to this polymer (see Figure 5). With addition of ZIL the polymers were obtained as flexible films. These films became clear with increasing ZIL content. The value of these films is based on the dissociation of LiTFSI upon adding ZIL to the polymer matrix. We have already reported that the mixture of ZIL and LiTFSI was obtained as viscous liquid, so that ZIL has an ability to promote dissociation of LiTFSI (28). This agrees with the findings of Davis et al. (35) and Shreeve et al (36). They confirmed that ZIL/acid mixtures act as an acidic catalyst for the esterification of alcohols and acetate derivatives. As acids they used acids HTFSI and CF_3SO_3H instead of LiTFSI, but found similar promotion of the reaction. Furthermore, the decomposition temperature (Td) of the ZIL/LiTFSI mixture was around 400 °C (37), whereas Td for pristine **D** is 330 °C. The Td of **D** increased by about 70 °C following LiTFSI addition. This implies that the imidazolium cation of ZIL interacts with the TFSI anion. This is supported by the fact that the Td of the mixture agrees with the value for common ILs containing TFSI anion (25).

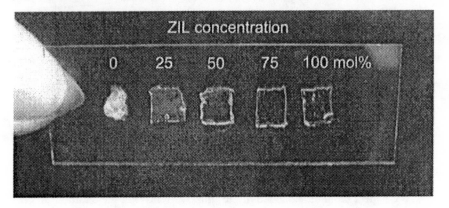

Figure 6. Photograph of solid polymer electrolytes composed of polymerized A with B_2 as a cross-linker, ZIL (D), and LiTFSI 50 mol%.

To investigate the effect of ZIL addition to the IL polymer, the degree of dissociation of LiTFSI was analyzed by Raman spectroscopy. Figure 7 shows Raman spectra of the IL polymer with 50 mol% LiTFSI and differing amounts of **D**. We have already confirmed that the band positions of the TFSI anion and of **D** are completely different. The IL polymer containing 50 mol% LiTFSI shows two bands, at 742 cm^{-1} and 746 cm^{-1}, in the absence of **D**. The band at 746 cm^{-1} disappeared upon adding **D**. The peak at 742 cm^{-1} shifted to 741 cm^{-1}. This implies that the fraction of free TFSI anions increased upon adding **D**. These results strongly suggest that **D** promotes the dissociation of LiTFSI in the IL polymer. It follows that the ionic conductivity and the lithium transference number (t_{Li+}) should improve with addition of **D**. Furthermore, mixtures of ZIL and LiTFSI are stable up to 400 °C. They can be used as non-volatile solvents for electrolyte materials. In fact the mixture of ZIL and LiTFSI showed high t_{Li+} exceeding 0.5 (38). We have not yet measured t_{Li+} for the IL polymer containing LiTFSI and ZIL; this will be reported elsewhere.

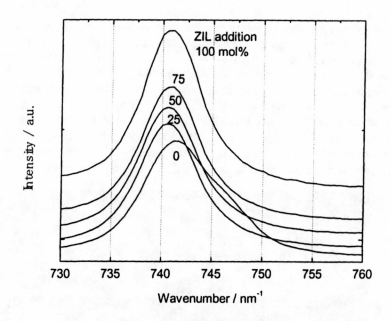

*Figure 7. Raman spectra of IL polymer A with B_2 and LiTFSI 50 mol%
containing various amounts of ZIL D.*

We expect the ionic conductivity to improve following the addition of ZIL, since ZIL promotes the dissociation of LiTFSI in the polymer matrix. Figure 8 shows the relation between the ZIL concentration and the ionic conductivity at 30 °C and Tg for the IL polymer **A** with **B$_2$** containing LiTFSI 50 mol% relative to the imidazolium unit. Addition of **D** clearly reduces the Tg of this system, following which further **D** gradually increases Tg. Since the addition of ZIL generates no carrier ions, mobile ions are decreased by adding ZILs to the polymer. This could explain the decrease in ionic conductivity upon adding ZILs. Excess ZIL would inhibit ion transport in the IL polymer. This agrees with our latest results that mixtures of LiTFSI and ZIL have maximum conductivity at the equimolar mixture (39).

The ionic conductivity of the IL polymer was not improved by the addition of ZIL. However, we have confirmed that ZIL promotes the dissociation of salt in the polymer matrix. We are currently trying to improve physical properties of these materials, especially Tm for ZIL and Tg for IL polymer. We are confident that materials having excellent properties will soon be obtained both as ZIL and IL polymer.

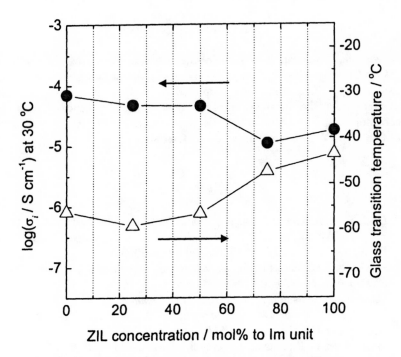

Figure 8. Effect of ZIL concentration on the ionic conductivity (•) and Tg (∆) for solid polymer electrolyte composed of polymerized A with B₂ as a cross-linker containing LiTFSI 50 mol% relative to the Im unit.

Conclusions

We have prepared IL polymers containing LiTFSI in various concentrations and investigated the ionic conductivities and dissociation behavior of the added salt. The ionic conductivity decreased monotonically with increasing LiTFSI concentration. Raman spectra revealed that free TFSI anions decreased monotonically with increasing LiTFSI concentration. ZIL was mixed with this polymer electrolyte as a non-volatile solvent that generated no carrier ions. The fraction of free TFSI anions was increased by the addition of ZIL when the LiTFSI concentration was 50 mol% relative to the imidazolium unit. We confirm that the ZIL is capable of promoting dissociation of LiTFSI in the IL polymer.

Acknowledgements

We thank Associate Prof. Dr. Nobuhumi Nakamura and Mr. Norio Suzuki (Tokyo University Agriculture & Technology) for useful advice on Raman spectroscopy. This study was supported by a Grant-in-Aid for Scientific Research from the Ministry of Education, Culture, Science, Sports and Technology, Japan (#14205136). The present study was carried out under the 21st Century COE program, Future Nano Materials.

References

1. (a) *Ionic Liquids-The Front and Future of Material Development-*; Ohno, H., Ed.; CMC: Tokyo, 2003 (b) *Ionic Liquids in Synthesis*; Wasserscheid, P.; Welton, T., Eds.; Wiley-VCH: Weinheim, 2003.
2. Nakagawa, H.; Izuchi, S.; Kuwana, K.; Nukuda, T.; Aihara, Y. *J. Electrochem. Soc.* **2003**, *150*, A695.
3. Sakaebe, H.; Matsumoto, H. *Electrochem. Commun.* **2003**, *5*, 594.
4. Hayashi, A.; Yoshizawa, M.; Angell, C. A.; Mizuno, F.; Minami, T.; Tatsumisago, M. *Electrochem. Solid-State Lett.* **2003**, *6*, E19.
5. MacFarlane, D. R.; Huang, J.; Forsyth, M. *Nature* **1999**, *402*, 792.
6. Koch, V. R.; Nanjundiah, C.; Appetecchi, G. B.; Scrosati, B. *J. Electrochem. Soc.* **1995**, *142*, L116.
7. Ogihara, W.; Yoshizawa, M.; Ohno, H. *Chem. Lett.* **2002**, 880.
8. Katayama, Y; Yukumoto, M; Miura, T. *J. Electrochem. Soc.* **2003**, *6*, A96.
9. Howlett, P. C.; MacFarlane, D. R.; Hollenkamp, A. F. *J. Power Sources* **2003**, *114*, 277.
10. Hu, Y.; Li, H.; Huang, X.; Chen, L. *Electrochem. Commun.* **2004**, *6*, 28.
11. Howlett, P. C.; MacFarlane, D. R.; Hollenkamp, A. F. *Electrochem. Solid-State Lett.* **2004**, *7*, A97.
12. Doyle, M.; Choi, S. K.; Proulx, G. *J. Electrochem. Soc.* **2000**, *147*, 34.
13. (a) Sun, J.; MacFarlane, D. R.; Forsyth, M. *Electrochim. Acta* **2001**, *46*, 1673. (b) Sun, J.; Jordan, L. R.; Forsyth, M.; MacFarlane, D. R. *Electrochim. Acta* **2001**, *46*, 1703.
14. (a) Noda, A.; Susan, M. A. B. H.; Kudo, K.; Mitsushima, S.; Hayamizu, K.; Watanabe, M. *J. Phys. Chem. B* **2003**, *107*, 4024. (b) Susan, M. A. B. H.; Noda, A.; Mitsushima, S.; Watanabe, M. *Chem. Commun.* **2003**, 938. (c) Susan, M. A. B. H.; Yoo, M.; Nakamoto, H.; Watanabe, M. *Chem. Lett.* **2003**, *32*, 836.

15. Yoshizawa, M.; Xu, W.; Angell, C. A. *J. Am. Chem. Soc.* **2003**, *125*, 15411.
16. Papageogiou, N.; Athanassov, Y.; Armand, M.; Bonhôte, P.; Pettersson, H.; Azam, A.; Grätzel, M. *J. Electrochem. Soc.* **1996**, *143*, 3099.
17. Kawano, R.; Watanabe, M. *Chem. Commun.* **2002**, 330.
18. Kubo, W.; Kitamura, T.; Hanabusa, K.; Wada, Y.; Yanagida, S. *Chem. Commun.* **2002**, 374.
19. Wang, P.; Zakeeruddin, S. M.; Exnar, I.; Grätzel, M. *Chem. Commun.* **2002**, 2972.
20. Matsumoto, H.; Matsuda, T.; Tsuda, T.; Hagiwara, R.; Ito, Y.; Miyazaki, Y. *Chem. Lett.* **2001**, 26.
21. Stathatos, E.; Lianos, P.; Zakeeruddin, S. M.; Liska, P.; Grätzel, M. *Chem. Mater.* **2003**, *15*, 1825.
22. Lewandowski, A.; Swiderska, A. *Solid State Ionics* **2003**, *161*, 243.
23. (a) Ue, M.; Takeda, M.; Takahashi, T.; Takehara, M. *Electrochem. Solid-State Lett.* **2002**, *5*, A119. (b) Ue, M.; Takeda, M.; Toriumi, A.; Kominato, A.; Hagiwara, R.; Ito, Y. *J. Electrochem. Soc.* **2003**, *150*, A499.
24. Stenger-Smith, J. D.; Webber, C. K.; Anderson, N.; Chafin, A. P.; Zong, K.; Reynolds, R. *J. Electrochem. Soc.* **2002**, *149*, A973.
25. McEwen, A. B.; Ngo, H. L.; LeCompte, K.; Goldman, J. L. *J. Electrochem. Soc.* **1999**, *146*, 1687.
26. McEwen, A. B.; McDevitt, S. F.; Koch, V. R. *J. Electrochem. Soc.* **1997**, *144*, L84.
27. Nanjundiah, C.; McDevitt, S. F.; Koch, V. R. *J. Electrochem. Soc.* **1997**, *144*, 3392.
28. (a) Yoshizawa, M.; Hirao, M.; Ito-Akita, K.; Ohno, H. *J. Mater. Chem.* **2001**, *11*, 1057. (b) Yoshizawa, M.; Narita, A.; Ohno, H. *Aust. J. Chem.* **2004**, *57*, 139.
29. (a) Yoshizawa, M.; Ohno, H. *Electrochim. Acta* **2001**, *46*, 1723. (b) Yoshizawa, M.; Ogihara, W.; Ohno, H. *Polym. Adv. Technol.* **2002**, *13*, 589. (c) Washiro, S.; Yoshizawa, M.; Nakajima, H.; Ohno, H. *Polymer* **2004**, *45*, 1577.
30. Ohno, H.; Inoue, Y.; Wang, P. *Solid State Ionics* **1993**, *62*, 257.
31. *Solid State Electrochemistry*; Bruce, P. G., Ed.; Cambridge University Press: Cambridge, 1995.
32. (a) Ohno, H.; Yoshizawa, M. *Solid State Ionics* **2002**, *154-155*, 303. (b) Hirao, M.; Sugimoto, H.; Ohno, H. *J. Electrochem. Soc.* **2000**, *147*, 4168. (c) Ohno, H.; Ogihara, W.; Yoshizawa, M. *Electrochim. Acta* in press.
33. (a) Rey, I.; Johansson, P.; Lindgren, J.; Lassegues, J. C.; Grondin, J.; Servant, L. *J. Phys. Chem. A* **1998**, *102*, 3249. (b) Rey, I.; Lassegues, J. C.; Grondin, J.; Servant, L. *Electrochim. Acta* **1998**, *43*, 1505.

34. Wang, Z.; Gao, W.; Huang, X.; Mo, Y.; Chen, L. *J. Raman Spectrosc.* **2001**, *32*, 900.
35. Cole, C. A.; Jensen, J. L.; Ntai, I.; Tran, K. L. T.; Weaver, K. J.; Forbes, D. C.; Davis, Jr., J. H. *J. Am. Chem. Soc.* **2002**, *124*, 5962.
36. Mirzaei, Y. R.; Xue, H.; Shreeve, J. M. *Inorg. Chem.* **2004**, *43*, 361.
37. Ohno, H.; Yoshizawa, M.; Ogihara, W. *Electrochim. Acta* **2003**, *48*, 2079.
38. Narita, A.; Yoshizawa, M.; Ohno, H. in preparation.
39. Yoshizawa, M.; Suzuki, N.; Nakamura, N.; Ohno, H. in preparation.

Chapter 7

Ionic Liquids as Alternative Plasticizers for Poly(vinyl chloride): Flexibility and Stability in Thermal, Leaching, and UV Environments

Mustafizur Rahman, Hugh W. Shoff, and Christopher S. Brazel

Department of Chemical and Biological Engineering, The University of Alabama, Tuscaloosa, AL 35487–0203

Ionic liquids (ILs) based on ammonium, imidazolium and phosphonium cations were proven to be good plasticizers for poly(vinyl chloride), PVC, offering similar or better properties compared to traditional plasticizers. Formulations containing as much as 50 wt % plasticizer were made, and tested for flexibility in comparison to traditional plasticizers based on phthalate, trimellitate, and citrate esters. The flexibilities of all plasticized samples were nearly equal, with slight variances in glass transition temperatures, depending primarily on the plasticizer content, but to a smaller extent on the chemical structure of the plasticizer. Because ILs are known for their wide liquid range and non-volatility, they can be used in plastics at higher temperatures than traditional plasticizers. The thermal stability of plasticized PVC was evaluated at moderate and high temperatures, as was each plasticizer's ability to reduce brittleness caused by ultraviolet (UV) light exposure. Solid-solid migration and solid-liquid leaching experiments were carried out to determine the suitability of IL plasticizers for the automotive and medical industries.

Background

Plasticizers are added to polymer formulations to reduce glass transition temperatures and impart flexibility to materials under conditions where they are normally used. Many traditional plasticizers are based on alkyl esters that often surround a ring structure and include phthalates, trimellitates, adipates, and, more recently, citrates. Di(2-ethyl hexyl) phthalate (DEHP), commonly known in industry as dioctyl phthalate, is the most widely-used plasticizer, as it is highly compatible with PVC and other polymers. DEHP accounts for nearly 50 % of plasticizers used in industry [1], and is suitable for a broad range of applications: consumer products (garden hoses, shower curtains), automotive parts, home construction materials, toys, and medical plastics. However, there remain a number of challenges in the plasticizer industry to develop new plasticized materials that have better performance in harsh environments, longer lifetimes, and reduced toxicity. A recent review paper [2] discusses some of the major approaches to developing these improved materials.

Among the approaches to develop novel flexible materials, ILs have many advantages for use as plasticizers. The low volatility and high temperature stability of many ILs make them useful for applications at elevated temperatures with minimal loss in flexibility and extended material lifetime. The cations and anions in ILs also have a strong affinity for each other, making plasticizer loss by liquid leaching, solid-solid migration, or evaporation much less likely compared to molecular plasticizers. Some of the potential advantages of using ILs as plasticizers are listed in Table 1.

The use of ILs in polymer systems has been explored on a number of fronts, but the majority of early papers in this field have focused on polymerization using ILs as solvents, with emphasis on the environmental aspects and ability to form new structures [3-5]. Attributes of these reactions, such as long free radical lifetimes, high molecular weight products, rapid polymerization, and the ability to form block copolymers without having to use tedious anionic or cationic polymerization schemes, have made a significant impact in the polymer field. However, industrial polymers made using ILs as environmentally-sound solvents have had considerably less success, as the separation step after the reaction has proven difficult and uneconomical. On the other hand, ILs can be formulated as plasticizers by at least three methods: direct compounding (using a high shear mixer, as most plasticizers are added), solvent casting, and polymerization using the plasticizer as the reaction solvent. Imidazolium-based ILs have been used effectively to plasticize PMMA, and with a higher degree of compatibility compared to DEHP [6]. After the experiments reported earlier, we sought to expand the range of ILs studied, focus on a more widely-used flexible plastic, PVC, and probe industrially-relevant properties including UV-stability and leaching. Results reflecting these new areas are presented here.

Table 1. Potential Advantages of ILs as Plasticizers

Good Low Temperature Lubricity
High Temperature Stability
Reduced Leaching in Aqueous and Lipophilic Solutions
Enhanced Stability to Ultraviolet Light
Reduced Flammability
Reduced Solid-Solid Migration
Reduced Evaporative Loss
High Degree of Compatibility in Polymer Formulations
Reduced Human Exposure

Objectives

The primary objective of this research was to determine the feasibility of using ILs to replace traditional plasticizers in PVC and extend the application range of flexible materials. More specifically, the experimental objectives were to:

- determine compatibility of phosphonium, ammonium and imidazolium ILs with PVC,
- measure thermal stability of plasticized PVC at high (> 200 °C) and moderate (~ 100 °C) temperatures,
- determine how the plasticizer molecular structure affects the flexibility of plasticized samples,
- measure any loss (or gain) in flexibility after exposure to ultraviolet light, and
- determine mass loss due to solid-solid migration and solid-liquid leaching.

Experimental

Polymers were plasticized with ILs and traditional plasticizers (see Figure 1). Traditional plasticizers included di(2-ethylhexyl phthalate (DEHP, Sigma Chemical Company, St. Louis, MO), trioctyl trimellitate (TOTM, Aldrich Chemical Company, Milwaukee, WI), and diisodecyl phthalate (DIDP, Fluka Chemie GmbH, Switzerland); all were used as received. Two citrate esters (Morflex, Inc., Greensboro, NC) that were developed for use in medical plastics were also studied. ILs studied included three that contained phosphonium cations ([thtdPh$^+$][deca$^-$], [thtdPh$^+$][Tf$_2$N$^-$], and [thtdPh$^+$][Cl$^-$]), one with an ammonium cation [tbam$^+$][doss$^-$], which is commercially sold by Sachem, Inc. as Terrasail^{TM} and two with an imidazolium cation. The phosphonium ILs were supplied by

TRADITIONAL PLASTICIZERS

Phthalates

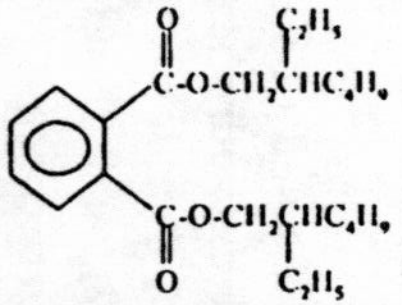

Diethylhexyl phthalate
DEHP

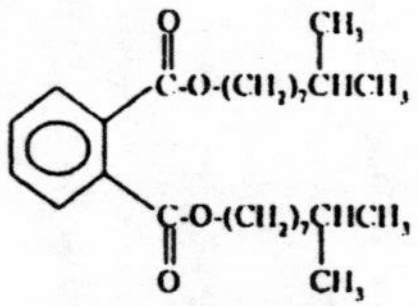

Diisodecyl phthalate
DIDP

Trimellitate

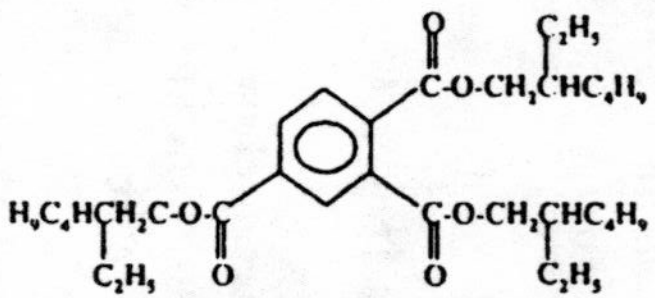

Trioctyl Trimellitate
TOTM

Citrates

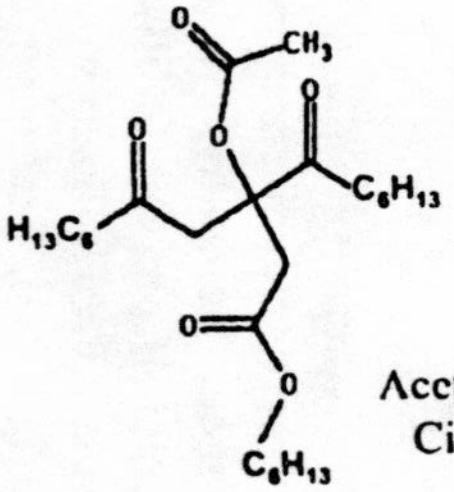

Acetyl tri-n-hexyl citrate
Citroflex® A-6 (A-6)

n-Butyryl tri-n-hexyl citrate
Citroflex® B-6 (B-6)

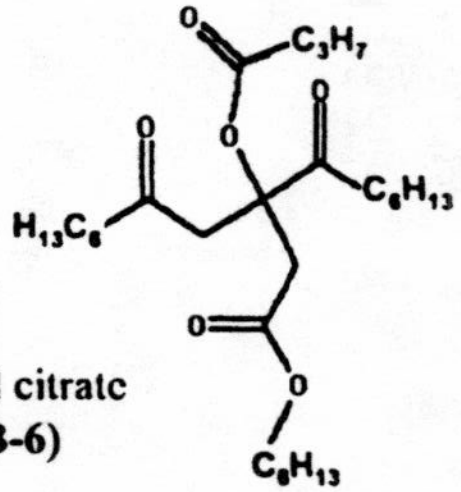

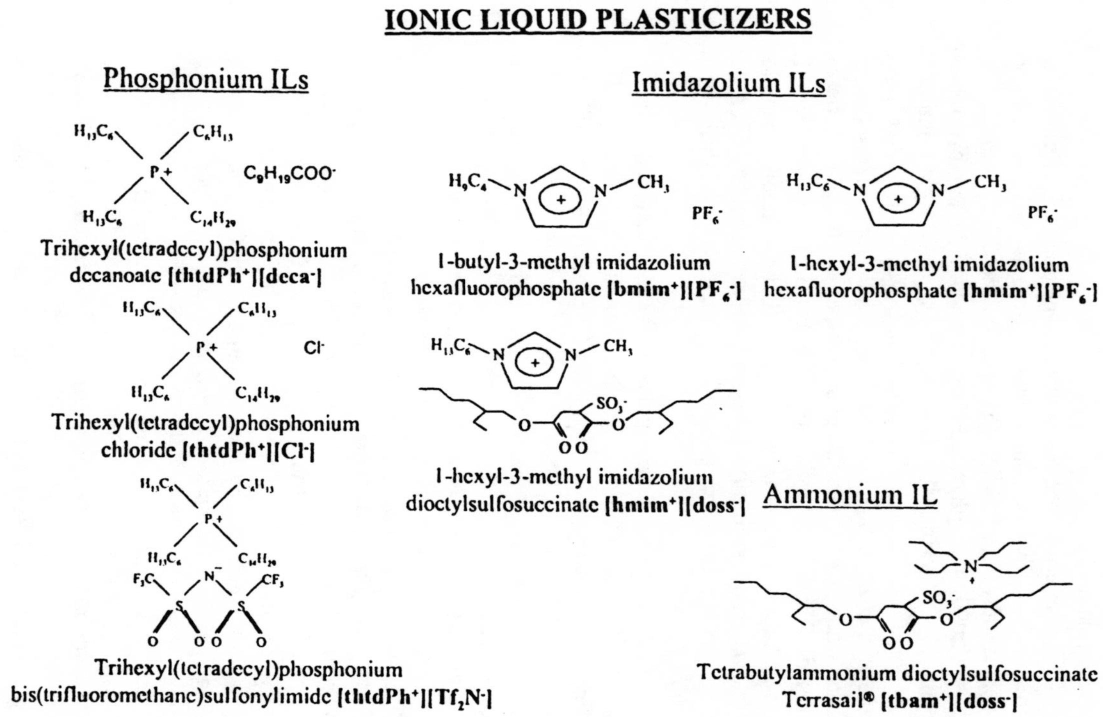

IONIC LIQUID PLASTICIZERS

Phosphonium ILs

Trihexyl(tetradecyl)phosphonium decanoate [thtdPh⁺][deca⁻]

Trihexyl(tetradecyl)phosphonium chloride [thtdPh⁺][Cl⁻]

Trihexyl(tetradecyl)phosphonium bis(trifluoromethane)sulfonylimide [thtdPh⁺][Tf₂N⁻]

Imidazolium ILs

1-butyl-3-methyl imidazolium hexafluorophosphate [bmim⁺][PF₆⁻]

1-hexyl-3-methyl imidazolium hexafluorophosphate [hmim⁺][PF₆⁻]

1-hexyl-3-methyl imidazolium dioctylsulfosuccinate [hmim⁺][doss⁻]

Ammonium IL

Tetrabutylammonium dioctylsulfosuccinate Terrasail® [tbam⁺][doss⁻]

Figure 1. Chemical Structures and Abbreviations for Traditional and Ionic Liquid Plasticizers Studied.

Cytec Industries, Inc. (Niagara Falls, Ontario, Canada); imidazolium ILs were synthesized by the Center for Green Manufacturing on the University of Alabama campus; the ammonium IL was supplied by Dr. Jim Davis of the University of South Alabama. The ILs were dewatered by placing under vacuum for 24 hours prior to formulation.

Plasticized PMMA samples were polymerized in the presence of plasticizer (acting as a reaction solvent that remained imbedded in the polymer network after the reaction), as described earlier [6]. Micronized PVC, supplied by Aldrich with M_n = 22,000, was plasticized by dissolving in tetrahydrofuran (THF, Acros Organics, Fair Lawn, NJ), mixing with a known weight fraction of plasticizer, and casting thin film samples by solvent evaporation. Plasticized samples were formulated at 20 wt % IL or traditional plasticizer as a baseline for analysis experiments, while several samples containing up to 50 wt % plasticizer were also formulated. Plasticized polymers were cut into different sample shapes: dogbone samples for Instron materials testing, thin rectangular strips for dynamic mechanical analysis and migration studies, and circular discs for leaching studies; small pieces (approximately 10 mg) were used for differential scanning calorimetry experiments.

Techniques used to analyze the plasticized polymer samples included scanning electron microscopy (SEM, Philips, Model XL-30, Eindhoven, The Netherlands), differential scanning calorimetry (DSC, Model 2920MDSC, TA Instruments, Newcastle, DE), an Instron® automated testing system (Model 5581, Canton, MA) and dynamic mechanical analysis (DMA, RSA II, Rheometrics Inc., Piscataway, NJ). DSC experiments were performed on plasticized samples to characterize glass transition temperatures; after a heating and cooling cycle to erase the thermal history of each sample, a ramp rate of 10 °C/min up to 150 °C was used. Elastic moduli were measured as a function of temperature by DMA, using a 0.01 %/min incremental strain rate during a temperature ramp from -50 to 50 °C for plasticized PVC. Thermogravimetric Analysis (TGA, Model 2950, TA Instruments, Newcastle, DE) was conducted using a thermal ramp rate of 10 °C/min from 30 to 350 °C. Long term thermal stability experiments were conducted by keeping samples at 105 °C in a heated oven, and measuring sample mass periodically over a 2-month period. An ultraviolet light source with a moving belt (Ultraviolet Systems and Equipment, Houston, TX) was used to control the exposure of plasticized samples to 254 nm UV light. Solid-solid migration experiments were conducted by placing rectangular-shaped plasticized PVC samples (surface dimensions: 33 x 7 mm; thickness: 1 mm) between two unplasticized PVC surfaces; the samples were clamped between glass slides and kept at 40 °C for 2 weeks, and mass losses from the plasticized samples were noted. Leaching experiments used tap water and saline solution (0.9 % NaCl) to extract plasticizers from 20 or 40 wt % plasticized PVC samples. The experiments were carried out for 7 days at 50 °C in a water bath shaken at 100 rpm, followed by water removal under vacuum. Sample masses were compared before and after leaching.

Results & Discussion

Uniformity

Imidazolium IL-plasticized PMMA samples have been shown to have similar if not superior performance to PMMA plasticized with DEHP [6-8]. At high plasticizer loading (30-50 wt% plasticizer), [bmim$^+$][PF$_6^-$] and [hmim$^+$][PF$_6^-$] formed uniform samples that continued to decrease the glass transition temperature, while DEHP phase separated from the host polymer at concentrations above 30 wt % in PMMA [6]. This earlier observation was confirmed using SEM (Figure 2). The DEHP-plasticized sample shows significant phase separation with only 30 wt % DEHP, while PMMA plasticized with as much as 50 wt % [hmim$^+$][PF$_6^-$] remains uniform, indicating greater compatibility between PMMA and the IL.

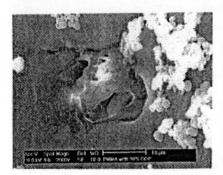 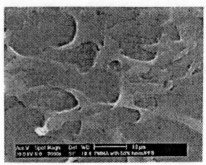

Figure 2. SEM Micrographs of PMMA Plasticized with (A) 30 wt % DEHP, and (B) 50 wt % [hmim$^+$][PF$_6^-$]

Flexibility

One of the most important characteristics of plasticizers is their ability to substantially lower the glass transition temperature to make the resulting materials flexible at the temperature of the desired application. Unplasticized PVC has a glass transition temperature of approximately 80 °C, but many uses for PVC are at or near room temperature. Plasticized PVC formulations often contain 50 wt % or higher amounts of plasticizer [9], so the experiments carried out provide an important screening of candidate IL plasticizers that can offer similar flexibility to PVC (Table 2). Here, three traditional plasticizers (one trimellitate- and two phthalate-based) were compounded into PVC and compared to unplasticized PVC and samples plasticized with various ILs. Each of the additives caused a notable decrease in T_g , as the lubricity between polymer chains was increased in every case. TOTM- and DEHP-plasticized PVC performed best, but two of the phosphonium-based ILs performed as well or better than traditional phthalate plasticizers DEHP and DIDP. The ammonium-based IL also lowered the T_g to near room temperature. Because the lubricity of plasticizers is based on polymer-plasticizer interaction as determined by their thermodynamic compatibilities, it is no surprise that the phosphonium IL containing a chloride anion performed well as a plasticizer.

A decrease in elastic modulus accompanies plasticization, with the modulus dropping by several orders of magnitude as the glass transition temperature is crossed (or when the plasticizer content is increased). Room temperature tensile moduli were determined for unplasticized PVC and PVC plasticized by both traditional and IL plasticizers (Table 3). The tensile modulus of PVC was reduced by approximately three orders of magnitude with the addition of 40 wt% plasticizer, with good reproducibility. Although DEHP and the Citroflex esters were better at reducing the tensile modulus of PVC, the plasticized samples tested at the same concentration of phosphonium- and imidazolium- based ILs still showed a significant reduction in elastic modulus. The higher values in this case are attributed to the higher viscosity (and molecular weight) of the ILs and potentially lower thermodynamic compatibility with PVC, reducing the effect of lubrication. Although [tbam$^+$][doss$^-$]-plasticized PVC had a higher tensile modulus, it was formulated at 20 wt% additive, whereas the other five samples

**Table 2. Glass Transition Temperatures of PVC Plasticized with 20 wt %
Traditional or IL Plasticizers. Where shown, errors represent the standard
deviation for three or more samples.**

Plasticizer	Glass Transition, T_g ($^{\circ}C$)
unplasticized PVC	80.1 ± 0.71
TOTM	10.1 ± 0.14
DEHP	14.2 ± 5.19
[thtdPh$^+$][Cl$^-$]	17.8 ± 0.92
[thtdPh$^+$][deca$^-$]	21.8 ± 0.71
DIDP	22.9 ± 2.75
[tbam$^+$][doss$^-$]	25.2 ± 0.64
[hmim$^+$][PF$_6^-$]	31.4
[bmim$^+$][PF$_6^-$]	35.8
[thtdPh$^+$][Tf$_2$N$^-$]	36.6 ± 0.55
[hmim$^+$][doss$^-$]	39.6 ± 3.39

Table 3. Tensile Moduli for PVC Plasticized with Traditional and IL Plasticizers, as Determined by Instron. Errors represent the standard deviation for three or more samples.

Plasticized PVC Sample	Tensile Modulus (MPa)	Strain at Break
Unplasticized PVC	3000 [10]	
40 wt % DEHP	0.800 ± 0.050	1.06 ± 0.020
40 wt % Citroflex A-6	1.96 ± 0.06	0.895 ± 0.233
40 wt % Citroflex B-6	0.766 ± 0.101	1.215 ± 0.035
40 wt % [thtdPh$^+$][Cl$^-$]	8.45 ± 0.16	0.850 ± 0.080
40 wt % [bmim$^+$][PF$_6^-$]	4.48 ± 0.14	1.470 ± 0.100
20 wt % [tbam$^+$][doss$^-$]	12.7 ± 0.5	0.510 ± 0.150

were formulated with 40 wt% plasticizer. In this case, a lack of available IL did not allow testing at the same concentration for adequate comparison purposes, but even with only 20 wt % [tbam$^+$][doss$^-$], the tensile modulus was reduced significantly. Additional experiments with [tbam$^+$][doss$^-$] are planned, as this IL shows promise not only for mechanical properties, but it is also lower in toxicity than imidazolium-based ILs, and has the potential to be produced more economically than other ILs. Data for elongation at break were taken for samples with initial length of one inch; in this test, [bmim$^+$][PF$_6^-$] outperformed all of the plasticizers tested.

Thermal Stability

Short- and long-term thermal stability studies show marked differences between the traditional phthalate and trimellitate plasticizers and selected ILs. In every case, exposing the plasticizers by themselves to a thermal ramp of 10 °C per minute caused significant evaporative losses with the traditional plasticizers, especially above 250 °C (Figure 3A). None of these plasticizers would be recommended for use above 200 °C, even for a brief time. The selected ILs showed much better thermal stability, with only small mass losses up to 300 °C

112

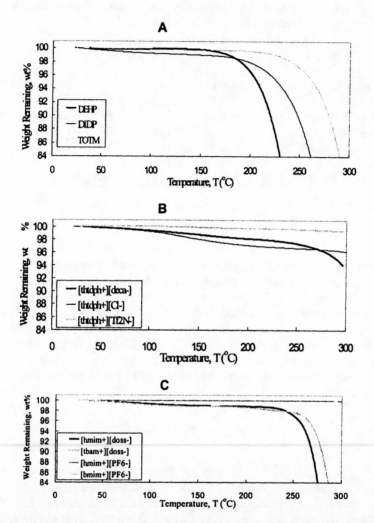

Figure 3. Mass Loss by Plasticizers during Temperature Ramp of 10 °C/min, as determined by TGA analysis. (A) Traditional Plasticizers DEHP, DIDP, and TOTM, (B) Phosphonium ILs, (C) Imidazolium and Ammonium ILs. In graph C, the curves for the imidazolium ILs are nearly identical.

Ultraviolet Stability

Intense ultraviolet light, or exposures over an extended period of time, often leads to rigid and brittle polymers. UV light can cause chain scission or branching in polymers, whether they are plasticized or not, and many plasticizers are also susceptible to UV degradation. Here, the PVC samples plasticized with ILs had the lowest increase in elastic modulus, with three of the IL-plasticized samples actually becoming softer after exposure to UV light (Table 4). The ideal case would be a system where the elastic modulus remains constant over the life of a product, but ILs show promise at preventing a loss in flexibility due to UV exposure. While these results show examples of how ILs can be used to prevent UV degradation, the mechanism has yet to be investigated. Zero mass loss was noted during the experiments, thus changes in flexibility are attributed to chemical changes in the PVC samples. The possibility that the ILs act to trap or scavenge free radicals is one plausible explanation, but will require further study.

Migration

Solid-solid migration of plasticizer results from the concentration gradient of internal plasticizer that causes slow diffusion through the void spaces in neighboring polymer parts. Because polymers are inherently porous (or at least have void spaces in non-crystalline regions), the mobility of plasticizer causes it to migrate during weeks to months to years of use. This problem is prevalent in automotive plastics, but can be minimized using surface treatments such as crosslinking to reduce plasticizer diffusion [11]. ILs provide another alternative: reducing diffusivity by using external plasticizers that are held together by weak ionic interactions. Here, several phosphonium-based ILs tested showed excellent migration resistance as less than 3 wt % of the initial plasticizer content was lost when plasticized PVC samples were sandwiched between two unplasticized PVC films. Compared to the traditional phthalate esters, of which up to 14 wt % migrated during the experiment, all of the ILs tested were significantly better suited for applications where solid-solid migration may cause plasticizer loss and brittleness.

Leaching

Leaching is one of the most difficult challenges facing plasticizer research: when plastics are subjected to liquids, diffusion from the surface and within the void space of the solid occurs at a greater magnitude than in migration. The probability and extent of leaching is largely dependent on the thermodynamic compatibility between the plasticizer and the leachate; thus the fluids that surround a plastic in its end-use dictate the appropriate methods to minimize leaching. This is important in many chemical processes, but one of the greatest challenges is developing plasticized materials for the medical industry, as IV bags and dialysis tubing use DEHP and other plasticizers to remain flexible. DEHP has been shown to leach into aqueous solutions including saline and human blood [12]. It is currently found at detectable levels in the blood supply [13]. Although DEHP has a relatively high LD_{50} value, it has been shown to bioaccumulate, so there is a real need to reduce leaching in medical devices.

Here, traditional phosphate and trimellitate ester plasticizers, as well as two medical-grade substitutes (Citroflex A-6 and B-6) were compared to several ILs by exposing plasticized PVC to leaching into tap water and saline solution (Table 6). Leaching into tap water (untreated city water supply) resulted in significant leaching for every material tested. Compared to DEHP, however, $[\text{thtdPh}^+][\text{Tf}_2\text{N}^-]$ and $[\text{tbam}^+][\text{doss}^-]$ showed reduced leaching at elevated temperatures. Leaching into unbuffered saline solution, while significantly lower than observed for the tap water leaching, was highly variable, most likely due to the counterdiffusion of NaCl into the PVC samples that occurred simultaneously to the removal of plasticizer. Since gravimetry was used for analysis, loss of plasticizer and uptake of NaCl cannot be decoupled, but work is underway to develop reliable techniques to analyze the leachate by chromatography to overcome this difficulty.

Conclusions

Ionic liquids show unique properties when used in polymeric systems. As plasticizers, they work as effectively as DEHP and other traditional plasticizers, but impart desirable properties to PMMA and PVC. The notable thermal stability of ILs compared to molecular solvents makes them particularly useful as plasticizers at high temperatures, as demonstrated in PMMA / imidazolium IL systems. However, the failure of phosphonium ILs to provide long term thermal

Table 4. Change in Elastic Moduli after Exposing Plasticized PVC to 254 nm Ultraviolet Light at 200 Wpi for Five Minutes. Moduli measured by DMA using an incremental strain rate of 0.01 %/min. Errors represent the standard deviation of three or more experiments.

Plasticized Sample	Initial Modulus at 25 °C [E_o] (GPa)	Modulus at 25 °C after UV Irradiation [E] (GPa)	Relative Change in Modulus, [E]/[E_o]
20% TOTM	0.552 ± 0.042	1.047 ± 0.016	1.90 ± 0.15
Bulk PVC (0%)	1.313 ± 0.072	2.330 ± 0.079	1.77 ± 0.11
20% DIDP	0.388 ± 0.026	0.539 ± 0.005	1.39 ± 0.09
20% DEHP	0.279 ± 0.012	0.379 ± 0.002	1.36 ± 0.06
20% [thtdPh$^+$][Tf$_2$N$^-$]	0.929 ± 0.073	1.158 ± 0.029	1.25 ± 0.10
20% [thtdPh$^+$][Cl$^-$]	0.855 ± 0.009	0.999 ± 0.005	1.17 ± 0.01
20% [thtdPh$^+$][deca$^-$]	0.582 ± 0.025	0.511 ± 0.007	0.88 ± 0.04
20% [hmim$^+$][doss$^-$]	1.238 ± 0.053	1.008 ± 0.012	0.81 ± 0.04
20% [tbam$^+$][doss$^-$]	0.755 ± 0.013	0.577 ± 0.016	0.76 ± 0.02

Table 5. Migration of Plasticizers from 20 wt % Plasticized PVC Samples While Surfaces Exposed to Unplasticized PVC for 2 weeks at 40 °C.

Plasticizer	Plasticizer Migrated to Unplasticized PVC (wt %)
[thtdPh$^+$][Cl$^-$]	0.00
[thtdPh$^+$][bta$^-$]	0.96
[thtdPh$^+$][deca$^-$]	2.74
[hmim$^+$][doss$^-$]	4.88
[tbam$^+$][doss$^-$]	5.99
TOTM	6.19
[bmim$^+$][PF$_6$$^-$]	7.67
DEHP	12.95
DIDP	13.99

Table 6. Mass Loss in 20 wt % Plasticized PVC Samples caused by Leaching in Tap Water or Saline Solution at 50 °C for 1 week. Where no values are entered, tests were not conducted. Errors represent the standard deviation of three samples.

Plasticizer	Tap Water (wt % leached)	Saline Solution (wt % leached)
DEHP	23.2	4.2 ± 0.40 / 12.5 ± 13.0[*]
TOTM	15.5	--
DIDP	23.8	--
Citroflex A-6	--	3.70 ± 16.8[*]
Citroflex B-6	--	1.68 ± 0.17[*]
[thtdPh$^+$][Cl$^-$]	28.7	4.98 ± 1.15[*]
[thtdPh$^+$][deca$^-$]	24.3	--
[thtdPh$^+$][Tf$_2$N$^-$]	14.7	--
[bmim$^+$][PF$_6^-$]	23.3	8.38 ± 0.38[*]
[tbam$^+$][doss$^-$]	17.0	[-7.12 ± 0.78][*,**]

* 40 wt % plasticizer was added to PVC

** an average net weight gain was recorded for this sample

stability to PVC at 105 °C raises questions about interactions between the IL and polymer that may be contributing to the degradation of PVC. The solvent-casting formulation techniques used here are useful at preparing uniform samples, with the IL or traditional plasticizer mixed throughout the polymer, but industrial manufacturing of flexible plastics usually involves high shear compounding mixers without the use of solvent. ILs have potential to overcome some of the ultraviolet light-induced brittleness that is common in phthalate-plasticized PVC, with flexibility retained in PVC plasticized with phosphonium-, ammonium-, and imidazolium-based ILs. Solid-solid migration was reduced significantly using ILs as plasticizers in the place of traditional phthalates. Finally, our initial research to study the leaching properties in plasticized PVC was inconclusive, as a better method to analyze small concentrations in aqueous solutions must be devised.

Acknowledgments

This research was supported by National Science Foundation grant NSF-CTS0086874 and a University of Alabama SOMED research grant. The authors also acknowledge Dr. David Nikles and Dr. Mark Weaver for the use of testing equipment. Ionic liquids were supplied by Cytec Industries, Inc. (phosphonium salts), Dr. Jim Davis of the University of South Alabama (ammonium salts), and Mr. Matthew Reichert of the UA Center for Green Manufacturing (imidazolium salts).

References

1. Murphy J, *Additives for Plastics Handbook*; Elsevier Science Ltd.: New York, NY, 2001.
2. Rahman, M.; Brazel, C.S. Review: An Assessment of Traditional Plasticizers and Research Trends for Development of Novel Plasticizers. *Prog. Polym. Sci.* **2004**, in review.
3. Harrisson, S.; Mackenzie, S.R.; Haddleton, D.M. Unprecedented Solvent-induced Acceleration of Free-Radical Propagation of Methyl Methacrylate in Ionic Liquids. *Chem. Commun.* **2002**, 2002, 2850-2851.
4. Zhang, H.; Hong, K.; Mays, J.W. Synthesis of Block Copolymers of Styrene and Methyl Methacrylate by Conventional Free Radical Polymerization in Room Temperature Ionic Liquids. *Macromolecules* **2002**, 35, 5738-5741.
5. Benton, M.G.; Brazel, C.S. An Investigation of the Degree and Rate of Polymerization of Poly(methyl methacrylate) in the Ionic Liquid 1-Butyl-3-Methylimidazolium Hexafluorophosphate. *Polymer International* **2004**, 53 1113-1117.
6. Scott, M.P.; Rahman, M.; Brazel, C.S. Application of Ionic Liquids as Low-Volatility Plasticizers for PMMA. *Eur. Polym. J.* **2003**, 39 1947-1953.
7. Scott, M.P.; Benton, M.G.; Rahman, M.; Brazel, C.S. Plasticizing Effects of Imidazolium Salts in PMMA: High Temperature Stable Flexible Engineering Materials, in Rogers, R.D., and K.R. Seddon, eds. *Ionic Liquids as Green Solvents: Progress and Prospects*, American Chemical Society Symposium Series: Washington, DC, 2003; 856 pp. 468-477.
8. Scott, M.P.; Brazel, C.S.; Benton, M.G.; Mays, J.W.; Holbrey, J.D.; Rogers, R.D. Application of Ionic Liquids as Plasticizers for Poly(methyl methacrylate)," *Chem. Commun.* **2002**, 2002 1370-1371.
9. Tickner, J.A.; Rossi M.; Haiama N.; Lappe M.; Hunt P. The use of di(2-ethylhexyl) phthalate in PVC medical devices: exposure, toxicity and alternatives. Lowell Center for Sustainable Production, Lowell, MA. 1999. <http://www.sustainableproduction.org/downloads/DEHP%20Full%20Text. pdf>
10. Collins, E.A.; Daniels, C.A.; Witenhader, D.E. Physical Constants of Poly(vinyl chloride) in Brandrup, J. and Immergut, E.H., eds. *Polymer Handbook*, Wiley Interscience: New York, NY, 1989; p. V-66.

11. Audic J.L.; Epaillard F.P.; Reyx D.; Brosse J.C. Cold plasma surface modification of conventionally and nonconventionally plasticized poly(vinyl chloride)-based flexible films: Global and specific migration of additives into isooctane. *J. Appl. Polym. Sci.* **2001**, 79, 1384-1393.

12. Jaeger, R.J.; Rubin R.J. Plasticizers from plastic devices: extraction, metabolism and accumulation by biological systems. *Science* **1970**, 170, 460.

13. Jaeger, R.J., Rubin R.J. Migration of phthalate ester plasticizer from poly(vinyl chloride) blood bags into stored human blood and its localization in human tissues. New. Engl. J. Med. **1972**, 287,1114-1118.

Chapter 8

Ion Gels Prepared by In Situ Radical Polymerization of Vinyl Monomers in Room Temperature Ionic Liquids as Novel Highly Conductive Polymer Electrolytes

Md. Abu Bin Hasan Susan[1,2], Akihiro Nada[1], and Masayoshi Watanabe[1,*]

[1]Department of Chemistry and Biotechnology, Yokohama National University, 79–5 Tokiwadai, Hodogaya-ku, Yokohama 240–8501, Japan
[2]Permanent address: Department of Chemistry, University of Dhaka, Dhaka 1000, Bangladesh

High ionic conductivity of ionic liquids has been exploited to realize novel polymer electrolytes. Certain vinyl monomers are found to be soluble in room temperature ionic liquids, and *in situ* free radical polymerization of the compatible monomers in the ionic liquids affords a novel series of polymer electrolytes. When the systems behave as a completely compatible binary system between the network polymer and an ionic liquid, we name these electrolytes as "ion gels". The obtained films of the ion gels are self-standing, flexible, and transparent, and the ionic conductivity at ambient temperature of the ion gels reaches to a value close to 10^{-2} Scm^{-1}. The ion-conduction behavior of the ion gels have been discussed with emphasis on the prospect as novel polymer electrolytes.

119

Polymer Electrolytes- A Brief Introduction

Polymer electrolytes, due to their unique properties, such as the ability to form thin films, flexibility, light weight, elasticity (plasticity), and transparency, have been recognized as the most viable electrolytes for potential application in solid-state electrochemical devices and sensors (1). For their practical applications in such devices, polymer electrolytes need essentially to have high ionic conductivity and wide potential window. These requirements have led to the surge of interest on the realization of electrochemically stable solid polymer electrolytes to ensure fast ion transport in the system resulting in the innovation of a wide variety of polymer electrolytes, which may be categorized conceptually in different classes (Figure 1).

Conventional ion-conducting polymers, such as polyether-based polymer electrolytes, are solid solutions of electrolyte salts in polymers (1). The ionic motion in these salt-in-polymer electrolytes is coupled with the local segmental motion, and since the generation of carrier ions occurs *via* the interaction of polymer segments and ions, increase in the carrier-ion density and mobility are inconsistent. An increase in ionic concentration thereby, causes an increase in the glass transition temperature (T_g). These facts are reflected by the appearance of a maximum in the ionic conductivity in polyethers with increasing salt concentration (Figure 1). To enhance the ambient temperature ionic conductivity in such salt-in-polymer electrolytes, the alternate approach has been the use of plasticizers. The ultimate materials are polymer gel electrolytes, where polymer electrolytes are swollen by organic electrolyte solutions.[1] These suffer from the drawbacks of flammability and volatility of the solvents used, which restricts their widespread use in large-scale devices. In a further development, the notion that a small amount of high-molecular weight polymer would suffice to give rubbery compliance with high ionic conductivity has been evident for polymer-in-salt systems (2). The electrolyte salts for such a polymer-in-salt system need to have some essential characteristics, such as low T_g and melting temperature (T_m), and to be (supercooled) liquids at ambient temperature while retaining high ionic conductivity. The number of carrier ions and their mobility for these polymer electrolytes increase with increasing electrolyte concentrations (Figure 1). Consequently, a high ionic conductivity that is not coupled with the segmental motion of the polymers is likely to be achieved.

Ionic liquids, also known as room temperature molten salts, due to their unique physicochemical properties, such as immeasurably low vapor pressure, high ionic conductivity, non flammability, and greater thermal and electrochemical stability (3-5), have been found suitable as electrolyte salts for polymer-in-salt systems. Ionic liquids meet all of the requirements of plasticizing salts and offer the potential for improved thermal and mechanical properties, and may expand the temperature range where flexible polymers can be used. In earlier articles, we have reported polymer-in-salt electrolytes, prepared by dissolving compatible polymers in chloroaluminate molten salts, which exhibit high ionic conductivity as well as rubbery electrolyte property (6). However, the chloroaluminate molten salts have quite high moisture sensitivity and the decomposed product by hydrolysis, HCl, is highly corrosive. Scott et al. have recently reported the ionic liquids based on imidazolium salts as excellent plasticizers for poly(methyl methacrylate) with improved thermal stability and

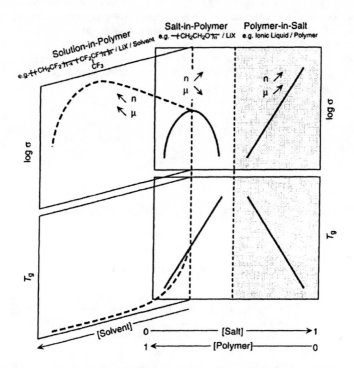

Figure 1. Conceptual classification of polymer electrolytes. n: carrier number, μ: ionic mobility, σ: ionic conductivity and T_g: glass transition temperature.

ability to significantly reduce the glass transition temperatures (7). Carlin and coworkers have also reported highly ion-conducting rubbery gel electrolytes from non-chloroaluminate ionic liquids and poly(vinylidene fluoride)-hexafluoropropylene copolymer [PVdF(HFP)] (8).

To bring perspective in this burgeoning field of polymer electrolytes, we therefore, aim at exploring the combination of air- and water- stable ionic liquids and polymers and understanding the ion-transport phenomenon. This paper deals with the preparation and characterization of polymer electrolytes based on ionic liquids, 1-ethyl-3-methylimidazolium tetrafluoroborate (EMIBF$_4$) and 1-butylpyridinium tetrafluoroborate (BPBF$_4$) (Scheme 1), and polymerization of vinyl monomers in the ionic liquids to present a new methodology affording highly conductive polymer electrolytes (9). We shall also report some of our results on polymer electrolytes based on the ionic liquid, 1-ethyl-3-methyl imidazoilium bis(trifluoromethane sulfonyl)imide (EMITFSI) (Scheme 1).

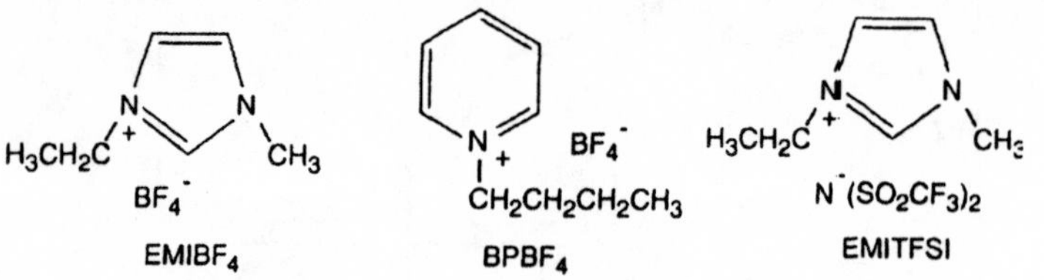

Scheme 1. Molecular structures of the room temperature ionic liquids.

Ion Gels: the Novel Polymer Electrolytes

The polymer-in-salt electrolytes have been made by dissolving compatible polymers in the ionic liquids, which afford high ionic conductivity as well as rubbery electrolyte properties. On the contrary, since a large number of vinyl monomers have been found to be soluble in ionic liquids, we have for the first time carried out *in situ* polymerization of the vinyl monomers in ionic liquids to afford novel polymer electrolytes (9). In fact, *in situ* free radical polymerization of compatible vinyl monomers in the presence of a small amount of a divinyl monomer in ionic liquids yields transparent, mechanically strong, and highly conductive polymer electrolyte films (9, 10). Like conventional polymer gels, the obtained polymer electrolytes are comprised of polymer networks and liquid, ionic liquid in our case. If the incorporation of ionic liquid into polymer networks affords completely compatible combination, we name the polymer gels obtained in this way as "ion gels" (10).

Ion gels are comprised of ionic liquids and network polymers, and retain the unique physicochemical properties of ionic liquids in the polymer electrolytes. If a compatible combination of ionic liquid with a thermally and electrochemically stable polymer network can be made, we will be able to use the novel polymer electrolytes for open-atmosphere and elevated-temperature operation. Furthermore, if task oriented properties, such as proton conduction, lithium ion conduction, and electron-transport are molecularly designed into ionic liquids, the scope and utility of ion gels will immediately expand to fuel cells (*10, 11*), lithium batteries (*12*), and solar cells (*13*).

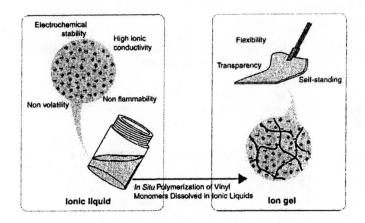

Figure 2. Ion gels obtained by in situ polymerization of vinyl monomers in room temperature ionic liquids. (10)

Ionic Liquids Used for the Preparation of Ion Gels

The room temperature ionic liquids, EMIBF$_4$, BPBF$_4$, and EMITFSI have been synthesized and used for the preparation of ion gels. Table I shows the thermal properties and ionic conductivities at 30 °C of the ionic liquids of our interest (*9*). In the DSC thermograms of EMIBF$_4$, BPBF$_4$, and EMITFSI neither exothermic nor endothermic peak has been observed except for the heat capacity change corresponding to the T_g during cooling from 100 °C to -150 °C. In the heating scans, the thermograms show heat capacity changes corresponding to the T_g, crystallization temperature (T_c) and T_m. Since the melting points of all of EMIBF$_4$, BPBF$_4$, and EMITFSI are below room temperature (Table I), we can confirm that these ionic liquids are room temperature ionic liquids. Below the melting points during the cooling scans, EMIBF$_4$, BPBF$_4$, and EMITFSI form supercooled liquids. We have already explored that the crystallization rates of these ionic liquids are very slow and that these supercooled liquids are fairly stable (9).

The ionic conductivities (σ) for EMIBF$_4$, BPBF$_4$, and EMITFSI at 30 °C are 20.0, 3.0, and 8.6 mScm^{-1}, respectively (Table I). The electrochemically stable potential windows of EMIBF$_4$ and BPBF$_4$, as determined by cyclic voltammetric measurements vs. Ag/AgCl at ambient temperature, are ca. 4.0 and ca. 3.4 V, respectively (9).

Table I. Thermal Properties and Ionic Conductivities at 30 °C for EMIBF$_4$ and BPBF$_4$, and EMITFSI

	$T_g/°C^a$	$T_c/°C^a$	$T_m/°C^a$	$T_d/°C^b$	σ at 30 °C / 10^{-3} Scm^{-1c}
EMIBF$_4$	-89.4	-50.5	14.6	391	20.0
BPBF$_4$	-66.7	-11.9	15.5	342	3.0
EMITFSI	-86.0	-92.0	-16.0	444	8.6

a Onset temperatures of a heat capacity change (T_g), an exothermic peak (T_c) and an endothermic peak (T_m) during heating scans from −150 °C by using differential scanning calorimetry. b Temperature of 10 % weight loss during heating scans from room temperature using thermo-gravimetry. c Ionic conductivity, σ, has been determined by complex impedance method in the frequency range of 5 Hz to 13 MHz at an AC amplitude of 10 mV.

Radical Polymerization of Vinyl Monomers in Ionic Liquids and Characterization of the Obtained Polymer Electrloytes

Certain vinyl monomers are soluble in ionic liquids and therefore, we have tried to polymerize common vinyl monomers using EMIBF$_4$ and BPBF$_4$ as the polymerization solvents. The compatibility of common vinyl monomers, methyl methacrylate (MMA), acrylonitrile (AN), vinyl acetate (VAc), styrene (St), and 2-hydroxyethyl methacrylate (HEMA) with EMIBF$_4$ and BPBF$_4$ have been checked by mixing equimolar amounts of EMIBF$_4$ or BPBF$_4$ and a monomer. Table II shows the compatibility of EMIBF$_4$ and BPBF$_4$ with the vinyl monomers. A circle in Table II means that the mixture is transparent and compatible, and a cross means that the mixture is not compatible and phase-separated. For example, in the case of BPBF$_4$, MMA, AN, VAc, and HEMA are compatible, whereas St is not compatible.

Table II. Compatibility[a] of EMIBF₄, and BPBF₄ with Vinyl Monomers and Their Polymers (14)

	$EMIBF_4$		$BPBF_4$	
	Monomer	*Polymer*	*Monomer*	*Polymer*
Methyl methacrylate	×	----	○	×
Acrylonitrile	○	×	○	×
Vinyl acetate	○	no polymerization	○	no polymerization
Styrene	×	—	×	—
2-Hydroxyethyl methacrylate	○	△	○	△

a ○: Transparent, △: Translucent, ×: Phase-separated *b* Ionic liquids and monomers have been mixed at equimolar ratio. *c* Polymerization has been carried out in the presence of 0.5 wt.% BPO at 80 °C for 12 h.

Source: Reproduced with permission from reference 14. Copyright 2000 Elsevier.

In the cases where either EMIBF₄ or BPBF₄ and the equimolar vinyl monomers are compatible, radical polymerization has been carried out by heating the solutions at 80 °C for 12 h in the presence of 0.5 wt% benzoyl peroxide (BPO) as an initiator. Prior to polymerization, the vinyl monomers have been purified by distillation and bubbled with nitrogen gas to ensure oxygen-free atmosphere. Interestingly, it has been found that the vinyl monomers are polymerized in EMIBF₄ or BPBF₄, except for VAc. This has been the first example of radical polymerization in ionic liquids (9). However, most of the resulting polymers are phase-separated from EMIBF₄ or BPBF₄. The compatibility of the polymers with EMIBF₄ and BPBF₄ are also shown in Table II. Under the experimental conditions, free radical polymerization reaction does not occur for VAc, possibly due to the inhibition caused by the trace amount of dissolved oxygen in the system.

Poly(HEMA) (PHEMA) shows comparatively good compatibility with both EMIBF₄ or BPBF₄ and therefore, PHEMA has been chosen as a matrix and the network polymer electrolytes based on HEMA and either EMIBF₄ or BPBF₄ have been prepared. Deoxygenated HEMA, ethylene glycol dimethacrylate (EGDMA, 2 mol% based on HEMA), and BPO (2 mol% based on HEMA) as an initiator are dissolved in the ionic liquids. The mixtures are spread between two glass plates separated by a polytetrafluoroethylene (Teflon®) spacer (0.5 mm in thickness), and *in situ* radical polymerization have been conducted at 80° C for 12 h. This gives transparent, self-standing, and mechanically strong polymer electrolyte films. The appearance of another ion gel example (EMITFSI/PMMA) is shown in Figure 3 as a photograph. By this procedure, solid polymer electrolytes with a variety of compositions have been obtained. The amount of unreacted monomers after the polymerization has been checked and has been found to be negligible for polymer electrolytes dissolved both in EMIBF₄ and BPBF₄. This indicates enhancement of the free radical polymerization (15) of vinyl monomers in ionic liquids. This is in good agreement with the corroborative study by Harrison et al., wherein by means of pulse laser polymerization technique, they have shown that the rate constant of MMA propagation increases considerably when polymerization is carried out in an ionic liquid (15a).

Figure 3. Photograph of a polymer electrolyte film of EMITFSI/PMMA ion gel.

As seen in the typical combinations of PHEMA and EMIBF$_4$ or BPBF$_4$, and PMMA and EMITFSI in the binary systems, the polymer electrolytes exhibit a single T_g in the range of the compositions, and the endothermic peak corresponding to the heat capacity change for melting point vanishes. Therefore, at all of the compositions of the ion gels, the system at temperatures higher than the T_g can be in the rubbery state. These suggest that the network polymer of PHEMA is compatible with dissolved ionic liquids. To clarify, the temperature dependence of dynamic mechanical properties for a BPBF$_4$ polymer electrolyte film ([BPBF$_4$]/[Monomer] = 4/6), has been shown in Figure 4, as a typical example (9). A large relaxation in tensile modulus (E') and a peak in tan δ are observed at ca. 20 °C. Although this temperature is close to the T_m of BPBF$_4$, the DSC thermogram does not exhibit any melting transition at this temperature. Thus, this relaxation temperature is assigned to the T_g of the polymer electrolyte. This is much higher than the T_g of BPBF$_4$ (-66.7 °C) and is much lower than the T_g of PHEMA (90 °C), indicating the plasticizing effect of the

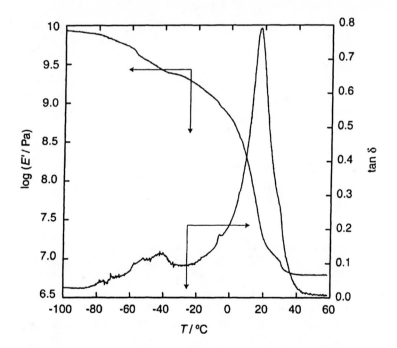

Figure 4. Storage modulus(E') and loss of tangent (tan δ) for HEMA network polymer with dissolved BPBF$_4$ ([BPBF$_4$/[HEMA] = 4/6] as a function of temperature. The sample has been rapidly cooled from ambient temperature to −150 °C, and the measurements have been carried out at the frequency of 10 Hz with the heating rate of 2 °C min^{-1}. (14)

(Reproduced with permission from reference 14. Copyright 2000 Elsevier.)

ionic liquid as well as compatibility between BPBF$_4$ and the HEMA network polymer.

Ionic Conductivity and Ion Transport Behavior in Ion Gels

The temperature dependencies of the ionic conductivity (σ) for the network polymer electrolytes based on PHEMA in EMIBF$_4$ and BPBF$_4$ along with those for bulk EMIBF$_4$ and BPBF$_4$ are depicted in Figures 5 and 6, respectively. The σ exhibits convex curved-profiles with change in temperatures and experimental data fit well with the Vogel-Tamman-Fulcher (VTF) equation[15]. The σ has been found to increase with increase in the content of EMIBF$_4$ or BPBF$_4$ in the ion gels. It is noteworthy that the remarkable decrease of the σ at the melting points of the ionic liquids is not observed in these results.

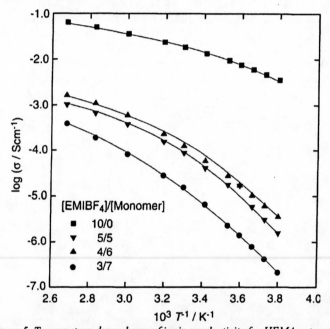

Figure 5. Temperature dependence of ionic conductivity for HEMA network polymers with dissolved EMIBF$_4$ and EMIBF$_4$ bulk. (14)

(Reproduced with permission from reference 14. Copyright 2000 Elsevier.)

The ionic conductivities of the EMIBF$_4$ electrolyte films are considerably lower than that of the ionic liquid itself (Figure 5). The electrolyte films have sufficient mechanical strength, and the phase-separation occurred at the higher

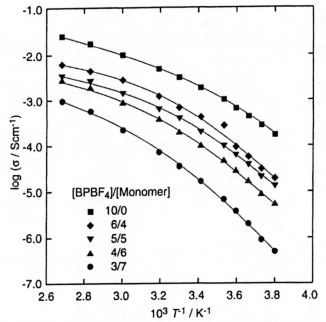

Figure 6. Temperature dependence of ionic conductivity for HEMA network polymers with dissolved BPBF₄ and BPBF₄ bulk. (14)

(Reproduced with permission from reference 14. Copyright 2000 Elsevier.)

compositions than $[EMIBF_4]/[Monomer] = 4/6$, giving translucent films. On the other hand, the compatibility of $BPBF_4$ with the HEMA network polymer is rather good, though the ionic conductivity of $BPBF_4$ itself is lower than that of $EMIBF_4$. The σ of the $[BPBF_4]/[Monomer] = 4/6$ polymer electrolyte reaches the value of ca. 10^{-3} Scm^{-1} at 30 °C, which is much higher than the values ever reported for conventional polyether electrolytes at ambient temperature. In fact, the films of ion gels based on PMMA network in EMITFSI, a perfectly compatible binary system, give the ambient temperature ionic conductivity in the order of as high as 10^{-2} Scm^{-1} *(17)*. The high ionic conductivity of the ion gels is due to the high self-dissociating and ion-transporting abilities of the constituent ionic liquids and to the decoupling of the ion transport from polymer segmental motion (*vide infra*).

To have a clear understanding of the phenomenological details of the ion transport in ion gels, the concept of "*coupling/decoupling*" and "*strong/fragile*" may be introduced here (Figure 7) *(18, 19)*. The *coupling* is a feature of a system, where the ion transport is frozen up in concurrence with the structure glass transition. In contrast in a *decoupling* system, conduction by the charge carriers is viable even below the T_g. The concept of *strong/fragile* demonstrates the difference in temperature dependence of conductivity from the VTF type (*fragile*) to Arrhenius type (*strong*).

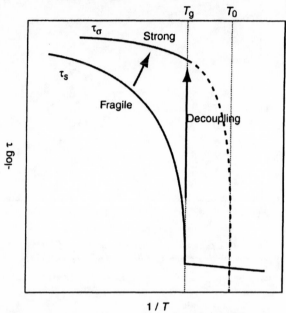

Figure 7. Temperature dependence of relaxation time in ion gels. τ_s : structure relaxation time, τ_σ :conduction relaxation time.

Our results on the EMITFSI/PMMA ion gels (*18*) show that with increasing polymer content, the difference between the T_g and T_0 (a parameter in the VTF equation) has been found to increase from a close value for bulk EMITFSI. At low salt concentrations, σ at the T_g shows a very high value ($10^{-7} - 10^{-8}$ Scm^{-1} order), which indicates that even if the systems are frozen, ionic transport still occurs. However, σ at the T_g decreases rapidly with increasing salt concentration. These support the transition from a *coupling* system to a *decoupling* one with decreasing salt concentration in the ion gels. In addition, at high salt concentrations, the σ exponentially decreases when approaching to the T_g, which indicates the systems to be a *fragile* one showing concurrence of structure relaxation and conduction relaxation. On the contrary, for high polymer content in the ion gels, the temperature dependence of ionic conductivity tends to change from VTF to Arrhenius behavior, and the σ can be measured even below the T_g. By increasing the polymer content in the ion gels, a transition from the *coupling/fragile* to a *decoupling/strong* system might occur in these systems. This is a very much fascinating and splendid phenomenon for molecular design of the polymer electrolytes of this variety.

Conclusion

The characteristics of ionic liquids have been successfully retained even with compatible polymers for the preparation of a novel series of polymer electrolytes. *In situ* free radical polymerization of suitable vinyl monomers in an ionic liquid affords ion gels that exhibit high ionic conductivities at room temperature, and possess sufficient mechanical strength, transparency, and flexibility, which are considered to be the basic requirements for polymer electrolytes. The ion transport behavior of the ion gels varies depending on the composition and structure of the constituents: the ionic liquid and polymer, and optimization of the factors controlling ion transport would enable tuning molecular engineering aspects of ion gels.

Acknowledgments

This research was supported in part by Grant-in-Aid for Scientific Research (#14350452 and #16205024) from the Japanese Ministry of Education, Science, Sports, and Culture and by Technology Research Grant Program from the NEDO of Japan. The authors also acknowledge kind experimental support from Taketo Kaneko and Shiro Seki. M.A.B.H.S. also acknowledges a post-doctoral fellowship from JSPS.

References

1. *Polymer Electrolyte Reviews 1 and 2*, Eds. MacCallum, J. R.; Vincent, C. A. Elsevier Appl. Sci. London, **1987** and **1989**.
2. Angell, C. A.; Liu, C.; Sanchez, E. *Nature* **1993**, *362*, 137.
3. Wilkes, J. S.; Zaworotko, M. J. *J. Chem. Soc., Chem. Commun.* **1992**, 965.
4. Noda, A.; Hayamizu, K.; Watanabe, M. *J. Phys. Chem. B.* **2001**, *105*, 4603.
5. *Ionic Liquids in Synthesis*, Eds. Wasserscheid, P.; Welton, T. Wiley-VCH Verlag, Weinheim, Germany, **2003**.
6. (a) Watanabe, M.; Yamada, S.; Sanui, K.; Ogata, N. *J. Chem. Soc., Chem. Commun.* **1993**, 929. (b) Watanabe, M.; Yamada, S.; Ogata, N.

132

Electrochim. Acta **1995**, *40*, 285. (c) Watanabe, M.; Mizumura, T. *Solid State Ionics* **1996**, *86-88*, 385.

7. (a) Scott M. P.; Brazel, C. S.; Benton, M. G.; Mays J. W.; Holbrey, J. D.; Rogers, R. D. *Chem. Commun.* **2002**, 1370. (b) Scott M. P.; Rahman, M.; Brazel, C. S. *Eur. Polym. J.* **2003**, *39*, 1947.

8. (a) Carlin, R. T.; Delong, H. C.; Fuller, J.; Trulove, P. C. *J. Electrochem. Soc.* **1994**, *141*, L73. (b) Carlin, R. T.; Fuller, J. *Chem. Commun.* **1997**, 1345. (c) Fuller, J.; Breda, A. C.; Carlin, R. T. *J. Electroanal. Chem.* **1998**, *45*, 29.

9. Noda, A.; Watanabe, M. *Electrochim. Acta* **2000**, *45*, 1265.

10. Susan, M. A. B. H.; Noda, A; Ishibashi, N.;Watanabe, M. in *Solid State Ionics: The Science and Technology of Ions in Motion* Eds. Chowdary, B.V.R. et al., World Scientific, Singapore, **2004**, pp. 899-910.

11. (a) Susan, M. A. B. H.; Noda, A; Mitsushima, S.;Watanabe, M. *Chem. Commun.* **2003**, 938. (b) Noda, A; Susan, M. A. B. H.; Kudo, K.; Mitsushima, S.; Hayamizu, K.;Watanabe, M. *J. Phys. Chem. B.* **2003**, *107*, 4024. (c) Susan, M. A. B. H.; Yoo, M.; Nakamoto, H.; Watanabe, M. *Chem. Lett.* **2003**, *32*, 836. (d) Susan, M. A. B. H.; Nakamoto, H.; Yoo, M.; Watanabe, M. *Trans. Mater. Res. Soc. Jpn.* **2004**, *29*, 1043.

12. Shoubukawa, H.; Tokuda, H.; Tabata, S.; Watanabe, M. *Electrochim. Acta* **2004**, in press.

13. Kawano, R.; Watanabe, M. *Chem. Commun.* **2003**, 330. (b) Kawano, R.; Matsui, H.; Matsuyama, C.; Sato, A.; Susan, M. A. B. H.; Tanabe, N.; Watanabe, M. *J. Photochem. Photobio. A, Chem.* **2004**, *164*, 87.

14. Reprinted from Noda, A.; Watanabe, M *Electrochim. Acta* **2000**, *8-9*, 1265, copyright 2000, Table 2, Figures 4-6, with permission from Elsevier.

15. Harrison, S.; MacKenzie, S. R.; Haddleton, D. M. *Chem. Commun.* **2002**, 2850. (b) Hong, K; Zhang, H.; Mays, J. W.; Visser, A. E.; Brazel, C. S.; Holbrey, J. D.; Reichert, W. M.; Rogers, R. D. *Chem. Commun.* **2002**, 1368. (c) Zhang, H.; Hong, K.; Mays, J. W. *Macromolecules* **2002**, *35*, 5738.

16. Vogel, H. *Phys. Z.* **1921**, *22*, 645. (b) Fulcher, G. S. *J. Am. Ceram. Soc.* **1923**, *8*, 339.

17. (a) Susan, M. A. B. H.; Kaneko, T.; Noda, A; Watanabe, M. to be submitted.

18. Seki, S.; Susan, M. A. B. H.; Kaneko, T.; Tokuda, H.; Watanabe, M. to be submitted.

19. (a) Angell, C. A. *Solid State Ionics* **1983**, *9-10*, 3. (b) Angell, C. A. *Solid State Ionics* **1986**, *18-19*, 72. (c) Bohmer, R.; Senapati, H.; Angell, C. A. *J. Non-Cryst. Solids* **1991**, *131-133*, 182. (d) Angell, C. A. *Science* **1995**, *267*, 1924. (e) Martinez, L.-M.; Angell, C. A. *Nature* **2001**, *410*, 663.

Chapter 9

Cross-Linked Polymers in Ionic Liquids: Ionic Liquids as Porogens

Peter Snedden[1], Andrew I. Cooper[1], Yaroslav Z. Khimyak[1],
Keith Scott[2], and Neil Winterton[1,*]

[1]Liverpool Centre for Materials and Catalysis, Department of Chemistry,
University of Liverpool, Liverpool L69 7ZD, United Kingdom
[2]Department of Chemical and Process Engineering, Merz Court,
University of Newcastle, Newcastle-upon-Tyne NE1 7RU, United Kingdom

Free-radical copolymerization of 4-vinylpyridine (VP) with
>5% divinylbenzene (DVB) in [omim]N(SO$_2$CF$_3$)$_2$, and
homopolymerizations of the cross-linking monomers DVB or
trimethylolpropane trimethacrylate (TRIM), led to gel-like
composite materials. VP copolymers with DVB showed a low
degree of permanent porosity in the dry state. However,
poly(DVB) and poly(TRIM) (after Soxhlet extraction) have
bulk densities, intrusion volumes, BET surface areas and
morphology (from SEM studies) which demonstrate the
porogenic character of ionic liquids. Poly(DVB) prepared in
ionic liquid was characterized by solid-state NMR and showed
a higher degree of cross-linking compared with material
prepared in toluene.

Introduction

Interest in ionic liquid-polymer composites (*1*) stems from the possibility that they may enable a serious constraint on the use of ionic liquids as reaction media in large-scale chemicals production to be circumvented. Compared with conventional molecular solvents, ionic liquids, generally, have significantly higher viscosity (*2*) and their distribution in polymer composites may improve contacting (*3*) of reactants and catalysts. Other approaches to limit such mass transfer constraints include the use of biphasic (*4*) or multiphasic (*5*) systems, immobilisation of ionic liquids on inorganic supports (*6*) or in inorganic (*7*) and organic (*8*) porous materials. Assembly of composites of polymer, ionic liquid and catalyst (in our case, for transition-metal catalysed hydrogenation of light alkenes (*8,9,10*)) may be achieved either by introducing a catalyst-ionic liquid solution into a polymeric membrane having permanent porosity or by *in situ* formation of a polymer gel, for example, by polymerization in/phase-separation from an ionic liquid. In either case, there must be compatibility between the components, the requisite degree of co-processibility in subsequent fabrication as well as essential mechanical and chemical robustness of the composite to the conditions of use. Fine control may, in principle, be achieved over the physical and chemical nature of the composite materials formed by *in situ* reaction-induced phase separation, offering a route to 'moulded' composites. In this case residual monomer may remain in the composite after reaction. However, since the ionic liquids chosen have essentially zero vapour pressure, it should be possible to remove traces of most monomers by evacuation at elevated temperatures.

To further these investigations, we have prepared and characterised a number of cross-linked polymeric materials using ionic liquids as the reaction medium (*1*), describing the first materials in which permanent porosity is achieved using the ionic liquid, [omim]Tf$_2$N, as porogenic solvent. (Conventional salts, such as [NH$_4$][HCO$_3$] and NaCl, may lead to pore formation (*11*) by gas evolution on heating or by leaching the solid.)

Polymers and polymerizations in ionic liquids

Polymerizations are among the earliest chemical reactions (*12*) to have been investigated in ionic liquids. While the interactions between polymers and ionic liquids are poorly understood, benefits have been reported (*13,14*) associated with increased rates of propagation and reduced rates of termination in [bmim]PF$_6$ compared with the molecular solvent benzene. The range of monomers polymerized in ionic liquids (*12,15*) is shown in Figure 1, and includes those with a polymerisable functionality incorporated into the ionic liquid itself. A wide spectrum of polymerization processes has been reported (*12,15*), including addition, condensation and ring-opening processes, involving electrode, enzymatic, Lewis acid, cation, co-ordination and free-radical, atom-transfer, reversible addition fragmentation chain-transfer mechanisms. The majority of these studies has focussed on homopolymerizations and some co-polymerizations, with very few (*16-18*) devoted to cross-linked polymers (Figure 2). None has reported the generation of porous materials in which the ionic liquid can act as a porogen (*19*).

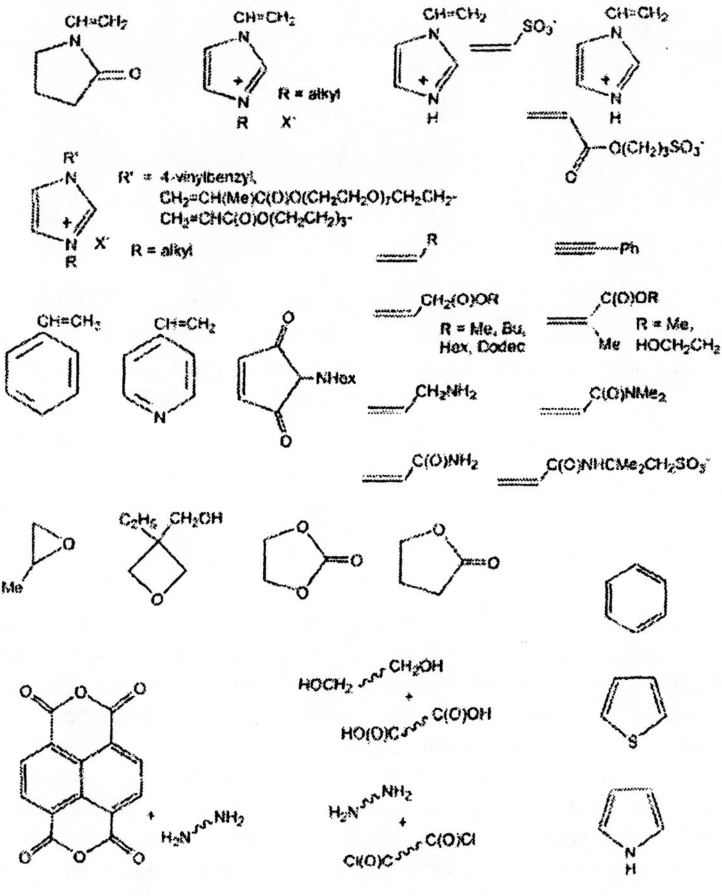

Figure 1. Monomers polymerized in ionic liquids

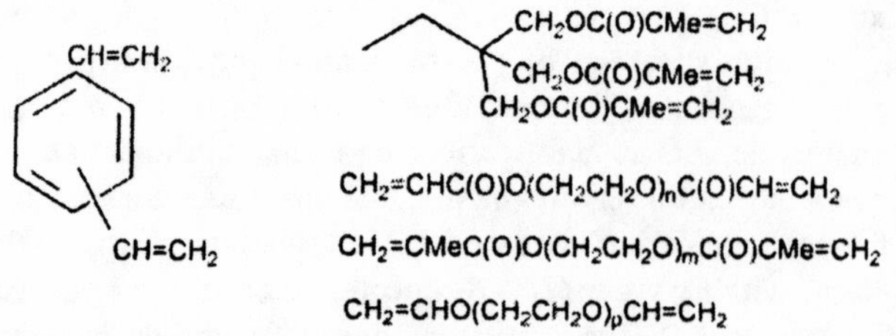

CH=CH$_2$

CH=CH$_2$

CH$_2$OC(O)CMe=CH$_2$
CH$_2$OC(O)CMe=CH$_2$
CH$_2$OC(O)CMe=CH$_2$

CH$_2$=CHC(O)O(CH$_2$CH$_2$O)$_n$C(O)CH=CH$_2$

CH$_2$=CMeC(O)O(CH$_2$CH$_2$O)$_m$C(O)CMe=CH$_2$

CH$_2$=CHO(CH$_2$CH$_2$O)$_p$CH=CH$_2$

Figure 2. Cross-linking monomers(1,16-18) used in polymerizations in ionic liquids

Experimental

Reagents

Divinylbenzene (DVB; tech., 55%, mixture of isomers), trimethylolpropane trimethacrylate (TRIM; tech.) and 4-vinylpyridine (VP; 95%) were used as supplied (Aldrich). α,α'-Azoisobutyronitrile (AIBN) (Fisher) was recrystallized from methanol prior to use. [omim]Tf$_2$N was prepared according to a published method (*20*).

Characterisation

GPC experiments were undertaken at 80 °C in DMF (0.01 M LiBr) using a Waters 150CV instrument fitted with a differential refractometer detector (columns: PLgel 2 × mixed bed–B, 30 cm, 10 μm; flow rate = 1.0 ml min^{-1}). Total intrusion volumes (V_{tot}), bulk densities (ρ_b), total pore areas (A_{tot}), median pore diameters (by volume, D_{mv}) and porosities (P) were measured using a Micromeritics AutoPore IV 9500 mercury intrusion porosimeter. Skeletal densities (ρ_s) were measured using a Micromeritics AccuPyc 1330 helium pycnometer. BET surface areas (SA$_{BET}$) were evaluated on a Micromeritics ASAP 2010 surface area and pore size analyzer; samples were degassed for 2 h at 110 °C prior to analysis. Scanning electron microscopy was performed on a Hitachi S-2460N scanning electron microscope. The samples were sputter-coated with approximately 10 nm gold before analysis.

^{1}H NMR spectra were recorded on a Bruker 400 MHz spectrometer (chemical shifts quoted in ppm relative to SiMe$_4$). Solid-state NMR spectra were acquired with a Bruker Avance DSX400 spectrometer operating at 400.16 MHz for ^{1}H and 100.63 MHz for ^{13}C with a MAS probehead using zirconia rotors 4 mm in diameter. The spinning rate was 7.0 kHz. ^{13}C chemical shifts are quoted in ppm relative to SiMe$_4$. ^{1}H–^{13}C CP/MAS NMR spectra with TPPM decoupling were acquired with ^{1}H π/2 pulse length 4.2 μs and pulse delay 10.0 s. The Hartmann-Hahn condition was set with hexamethylbenzene. The contact times were varied in the range of 0.02-12.0 ms. The dipolar dephasing spectra were recorded using contact time of 1.0 ms and dephasing time of 5, 10 and 20 μs. ^{13}C {^{1}H} MAS NMR spectra with TPPM decoupling were acquired with ^{13}C π/2 pulse length 4.5 μs and 120.0 s pulse delay to ensure full relaxation.

Procedures for polymerizations and polymer purification.

A degassed solution of monomer (VP (+ DVB (0, 1, 5 or 30 mol%)), DVB or TRIM; 29 v/v%), AIBN (1 mol%) and solvent ([omim]Tf$_2$N or toluene) in a Schlenk tube was stirred under vacuum at 80 °C for 4 h. Linear polymers were purified by repeated re-precipitation. Cross-linked polymers were purified by Soxhlet extraction (refluxing MeOH, 72 h). Purified polymers were dried in vacuo (50 °C) to constant weight.

Polymerization of VP and DVB (0, 5, 30 mol%). All reactions were carried out in [omim]Tf$_2$N. Poly(VP) was triturated 3 times with EtOAc then reprecipitated 5 times from CH$_2$Cl$_2$–EtOAc, giving a white solid, M_w = 71,000 g mol^{-1}, M_n = 36,000 g mol^{-1}, M_w/M_n = 2.0 (data averaged over two runs); δ_H (CDCl$_3$, 400 MHz) 1.44 (m, backbone; sh at 1.55), 1.84 (m, backbone; sh at 1.90), 6.38 (m, 2H, ArH) and 8.32 (m, 2H, ArH). DVB-cross-linked poly(VP)–[omim]Tf$_2$N composite material **A** (with 5 mol% DVB) is a transparent orange gel; **B** (with 30 mol% DVB) is a pale flesh-coloured waxy solid. DVB-cross-linked poly(VP)s were isolated after extraction: **A'** (5 mol% DVB), an orange solid; ~100% conversion and **B'** (30 mol% DVB), a flesh-coloured solid; 89% conversion.

Polymerization of DVB (55 mol%). Reactions were carried out in [omim]Tf$_2$N (**C**) or toluene (**E**). Poly(DVB)–[omim]Tf$_2$N composite material: **C**, white chalky solid. Poly(DVB)s: **C'** (after Soxhlet extraction of **C**), white solid (90% conversion); **E**, white solid (67% conversion). Sulfur content **C'** 0.16, 0.14 %; **E** < 0.01 % (*4*).

Polymerization of TRIM (100 mol%). Reactions were carried out in [omim]Tf$_2$N (**D**) or toluene (**G**). Poly(TRIM)–[omim]Tf$_2$N composite material: **D**, pale flesh-coloured waxy solid. Poly(TRIM)s: **D'** (after Soxhlet extraction of **D**), white solid (~100% conversion); **G**, white solid (97% conversion). Sulfur content **D'** 10.53, 10.62 %; **G** < 0.01 % (*4*).

Results and Discussion

Polymer and Composite Preparation

Polymerization of VP in [omim]Tf$_2$N led to a homogeneous viscous solution of a linear polymer with a weight-average molecular weight of 71,500 g mol^{-1} from which poly(4-vinylpyridine) was isolated as a white solid. The detection of residues of [omim]Tf$_2$N in the ^{1}H NMR of this material, despite repeated re-precipitations, confirms the difficulty (also noted by others (18)) that may exist in removing all traces of these non-volatile solvents from polymeric materials. Co-polymerization in [omim]Tf$_2$N of VP with 5 mol% DVB as cross-linker gave a transparent poly(4-vinylpyridine)-[omim]Tf$_2$N composite gel (A). An opaque gel (B) was formed with 30 mol% DVB. Both of these gels were completely insoluble in all organic solvents tested. Soxhlet extraction (methanol, 72 h) gave brittle orange solids (A' and B', respectively) with gel fractions of 100%.

Opaque white solid polymer–ionic liquid composite materials were also produced by free-radical polymerization of neat 55% DVB (C) and 100% trimethylolpropane trimethacrylate (TRIM) (D). Exhaustive extraction of C and D gave C' and D', respectively. Poly(DVB) E and poly(TRIM) G were prepared in toluene.

Composite and Polymer Characterisation

Importantly, the liquid phase of composite materials based on ionic liquids has essentially zero vapour pressure, allowing characteristics that usually require 'dry' samples, such as thermogravimetry, to be examined. Thermogravimetric analysis of the polymer–ionic liquid *composite* material, C and the isolated *polymer*, C' allows decomposition of the polymer, decomposition of the ionic liquid and loss of any volatiles associated with unreacted monomer to be assessed on related samples (1). polymer–IL composites were found to be somewhat more thermally stable than the respective polymer after extraction of the ionic liquid. We suspect that this may be associated with lower oxygen solubility in ionic liquids (21).

Table 1. Cross-linked polymer–ionic liquid composites (A-D) and corresponding cross-linked polymers (A'-D'; E, G)

Composite or Polymer	Monomer	X-linker (mol%)	Solvent	Appearance	ρ_s (g ml^{-1})^a (T/ °C)
A	VP	5^b	[omim]NTf$_2$	Gel	1.291 (26.1)
B	VP	30^b	[omim]NTf$_2$	Waxy solid	1.301 (24.6)
C	DVB	55	[omim]NTf$_2$	Chalky solid	1.282 (25.8)
D	TRIM	100	[omim]NTf$_2$	Waxy solid	1.327 (25.9)
A'	VP	5	[omim]NTf$_2$	Orange solid	1.278 (24.6)
B'	VP	30	[omim]NTf$_2$	Flesh coloured solid	1.141 (26.5)
C'	DVB	55	[omim]NTf$_2$	White solid	1.126 (24.7)
E	DVB	55	Toluene	White solid	1.080 (25.4)
D'	TRIM	100	[omim]NTf$_2$	White solid	1.264 (26.7)
G	TRIM	100	Toluene	White solid	1.245 (26.7)

^{a} Skeletal density (from helium pycnometry). ^{b} DVB (55%; remainder composed of a mixture of 3- and 4-ethylvinylbenzene).

Data from the analysis of pore structures by N_2 sorption–desorption, mercury intrusion porosimetry and helium pycnometry are shown in Table 2.
A' (from VP + 5% DVB) and B' (from VP + 30% DVB) have high bulk densities (>0.75 g cm^{-3}), low intrusion volumes (<0.3 g cm^{-3}) and low BET surface areas (< 2 m^2 g^{-1}), characteristics associated with a low degree of permanent porosity in the dry state. In addition, they show significant weight loss and pronounced shrinkage after extraction of the ionic liquid. A' and B' have some of the properties of gel-type resins (19) in that they swell in the presence of [omim]Tf$_2$N and deswell after Soxhlet extraction. (Linear (i.e., non-cross-linked) poly(VP) has some solubility in [omim]Tf$_2$N (1).) The degree of cross-linking in these materials appears to be insufficiently high to support the formation of permanent porosity (22,23).

Some interesting differences are seen when DVB and TRIM are polymerized alone. For instance, the degree of ionic liquid retention is significantly different for C' and D'. Sulfur analysis shows that very little ionic liquid is retained in poly(DVB) (C') following Soxhlet extraction of the IL-poly(DVB) composite, C, whereas poly(TRIM) retained significant amounts of [omim]Tf$_2$N, even after exhaustive extraction. In contrast to A' and B', both C', synthesized from DVB and D', prepared from TRIM, were found to have relatively low bulk densities (< 0.6 g cm^{-3}), higher intrusion volumes (> 0.9 cm^3 g^{-1}) and much higher BET surface areas (70–370 m^2 g^{-1}).

Table 2. Cross-linked polymers [a]

Material	Yield[b] (%)	ρ_b [c] (g ml⁻¹)	SA_{BET} [d] m² g⁻¹	V_{tot} [c] ml g⁻¹	A_{tot} [c] m² g⁻¹	D_{mv} [c] μm	P [c] (%)
A'	~100	1.050	1.6	0.13 [g]	-	-	<10 [e]
B'	89	0.768	1.7	0.26 [g]	-	-	<10 [e]
C'	90	0.242	73.2	2.62	53.6	0.57	63
E	67	0.788	321	0.17	19.1 [f]	0.056	14
G	59	0.741	4.5	0.26	66.3	0.033	19
D'	~100	0.559	365	0.89	170 [f]	0.040	50

[a] Polymerization conditions: monomer (1 vol), [omim]Tf₂N (2.5 vol), AIBN, 80 °C, 4 h. [b] After Soxhlet-extraction (refluxing MeOH, 72 h). [c] From mercury intrusion porosimetry: ρ_b, bulk density (at 1 psia); V_{tot}, total intrusion volume; A_{tot}, total pore area; D_{mv}, median pore diameter by volume; P, porosity. [d] BET surface area. [e] Most apparent porosity can be ascribed to interparticulate filling at low mercury pressure. [f] BET surface area suggests significant proportion of pores <7 nm that are not detected by mercury intrusion porosimetry.

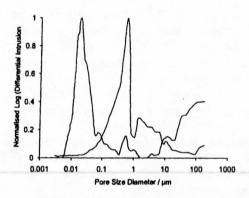

Figure 3. Pore-size distributions for Soxhlet extracted poly(divinylbenzene)s polymerized in [omim]NTf₂ (C') (right peak) and toluene (E) (left peak).
(Reproduced from reference 1. Copyright 2003 American Chemical Society.)

Furthermore, polymers **C'** and **D'**, prepared using [omim]Tf₂N as the porogen, were quite different from the control polymers (**E** and **G**) synthesized using toluene as the porogen. In the case of **C'**, the average pore size of the material was much greater than the equivalent polymer synthesized using toluene (**E**) (see Figure 3). In addition, the surface area for **C'** (73.2 m² g⁻¹) was much lower than for **E** (321 m² g⁻¹). The morphology of the porous materials was

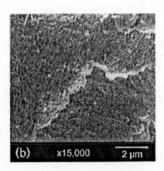

Figure 4. Scanning electron micrographs (×15,000) of Soxhlet-extracted poly(divinylbenzene)s (55 mol%); polymerization in (a) [omim]Tf₂N (C') and (b) toluene (E).

(Reproduced from reference 1. Copyright 2003 American Chemical Society.)

investigated by electron microscopy, **C'** displaying a much coarser structure (Figure 4a) than that observed for the equivalent polymer synthesized using toluene as the porogen (**E**, Figure 4b).

Solid-State NMR studies of poly(DVB)

The ^{13}C {^{1}H} magic-angle spinning (MAS) NMR spectra of poly(DVB) synthesised in [omim]Tf₂N (**C'**) and toluene (**E**) show all characteristic peaks typical of cross-linked DVB systems (*22,23*). The assignments of the peaks (Figures 5 and 6) were confirmed by ^{13}C {^{1}H} dipolar dephasing spectra (*24*). The key feature is the presence of resonances at *ca.* 138 and 113 ppm, attributable to the methine and methylene groups of unreacted vinyls, respectively. Peaks at *ca.* 29.5 and 16.0 ppm, observed for both **C'** and **E**, can be assigned to ethyl groups attached to the aromatic residues. The latter probably arise from the ethylstyrene impurities present in technical grade DVB.

The analysis of the ^{13}C {^{1}H} MAS and ^{1}H-^{13}C cross-polarisation/MAS (CP/MAS) spectra for **C'** and **E** has been used to estimate the relative degree of cross-linking. The spectra of DVB polymerized in toluene show higher population of resonances at 138 and 113 ppm compared to the spectra of poly(DVB) synthesised in ionic liquid. Quantitative analysis of the degree of cross-linking is hampered by the fact that significant spinning sidebands are observed in both ^{13}C {^{1}H} MAS and ^{1}H-^{13}C CP/MAS spectra even at 7.0 kHz.

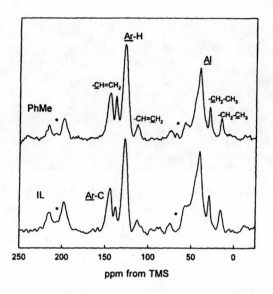

Figure 5. ^{13}C *{^{1}H} MAS NMR spectra of poly(DVB) in [omim] Tf₂N or toluene.*

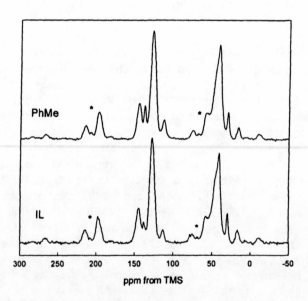

Figure 6. ^{1}H-^{13}C *CP/MAS NMR spectra of poly(DVB) in[omim]Tf₂N (IL) or toluene (contact time 1 ms).*

The spinning sidebands of the Ar-H carbons (128 ppm) partially overlap with the reconances attributable to the aliphatic carbons in the main polymer chains. The ratios of ArH/$\underline{C}$H=CH$_2$ and ArH/CH=$\underline{C}$H$_2$ populations were determined using deconvolution of the main resonances and considering the first order spinning sidebands (Table 3). In both cases, these ratios are much higher for the polymer synthesised in the presence of the ionic liquid, indicating higher degree of cross-linking.

Table 3. ArH/-CH=CH$_2$ ratios determined from
^{13}C {^{1}H} MAS NMR spectra

Sample	ArH/-$\underline{C}$H=CH$_2$	ArH/-CH=$\underline{C}$H$_2$
Poly(DVB) E	7.4	13.8
Poly(DVB) C′	16.1	19.8

The different polymer units show different ^{1}H-^{13}C CP/MAS kinetics. Cross polarisation relies on the transfer of magnetisation from the abundant ^{1}H system to the dilute ^{13}C atoms *via* heteronuclear dipolar coupling (*25*). The efficiency of CP depends both on the distance between ^{1}H and ^{13}C and the mobility of the CH–units. This explains the non-quantitative character of the CP/MAS spectra with the lines corresponding to rigid and proton-attached carbons exhibiting much faster CP-kinetics compared to either mobile -CH$_3$ groups or quaternary carbons. Analysis of the ^{1}H-^{13}C CP/MAS dynamics in **E** synthesised in the presence of toluene shows that the –CH=CH$_2$ groups are more mobile than Ar-H, Ar-C or aliphatic carbons (the T_{CH} and $T_{1\rho H}$ times derived from the CP-kinetics curves are much longer for –$\underline{C}$H=CH$_2$ and –CH=$\underline{C}$H$_2$ carbons). Detailed study of the influence of the ionic liquid on the mobility of the DVB is in progress.

Characterisation of poly(TRIM)

The solid-state NMR of poly(TRIM), **D′**, is still under investigation and will be reported in detail elsewhere. However, the spectra confirm that **D′** retains very substantial quantities of [omim]Tf$_2$N, despite lengthy Soxhlet extraction, in contrast to poly(DVB), **C′**.

The pore structure of **D′** was also analyzed by N$_2$ sorption–desorption, by mercury intrusion porosimetry and by helium pycnometry (Table 2), permitting the comparison of **D′** with the analogous material, **G**, prepared in toluene. In contrast to the comparison between **C′** and **E**, **D′**, prepared from TRIM in

144

[omim]Tf₂N, exhibits properties that are very similar to the equivalent material, **G**, synthesized in toluene. Both **D'** and **G** have high surface areas (~370 m^2 g^{-1}) and similar average pore sizes (0.040 and 0.029 μm, respectively) (Figure 7). **D'** and **G** (Figure 8a and 8b, respectively) also show very similar morphologies.

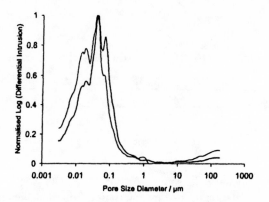

Figure 7. Pore size distribution for poly(TRIM)polymerized in [omim]Tf₂N (D')
(bold line) or in toluene (G)(faint line)
(Reproduced from reference 1. Copyright 2003 American Chemical Society.)

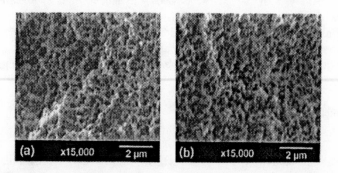

Figure 8. Scanning electron micrograph (×15,000) of Soxhlet-extracted
poly(trimethylolpropane trimethacrylate)s; polymerization in (a) [omim]Tf₂N
(D') and (b) toluene (G).

(Reproduced from reference 1. Copyright 2003 American Chemical Society.)

Porogenic behaviour

It is known that the polymerization medium affects the processes of phase separation, nucleation, aggregation, and pore formation (*20*), so the differences revealed, particularly between poly(DVB) **C′** and **E** may arise from the sensitivity of these systems to the nature of the porogenic solvent. Similar effects in conventional organic solvent porogens have been documented. For example, a dramatic porogen effect on the structure of porous copolymer monoliths synthesized from ethylene glycol dimethacrylate (EGDMA), butyl methacrylate, and 2-acrylamido-2-methyl-1-propanesulfonic acid (AMPS) has been reported by Peters *et al.* (*26*) who used a ternary porogen mixture (10 wt% water and 90 wt% of a mixture of 1-propanol and 1,4-butanediol). The average pore size in the monoliths was *extremely* sensitive to the porogenic solvent composition, decreasing from ~5000 nm to just 150 nm as the proportion of propanol was increased from 55 wt% to 63 wt% (*26*). It was suggested that this arose from the degree of solvation for the AMPS moieties, since the effect was more pronounced at higher AMPS incorporations. Similarly, the average pore size in EGDMA/glycidyl methacrylate copolymer monoliths was controlled by using a cyclohexanol/dodecanol porogen mixture and varying the ratio of the two alcohols (*27*). Xie *et al.* studied a homologous series of alcohols (methanol – dodecanol) mixed with dimethylsulfoxide as porogens for the preparation of poly(acrylamide-*co*-methylenebisacrylamide) monoliths and found that the average pore size increased significantly with increasing chain length of the porogenic alcohol (*28*). Each of these studies illustrates the sensitivity of the structure of highly cross-linked, permanently porous polyacrylates and polymethacrylates to the precise nature of the porogenic diluent. That the DVB system interacts quite differently with [omim]Tf$_2$N and toluene, while the TRIM system gives rise to similar materials of similar morphology, is further illustrative of this poorly understood phenomenon. Further investigations are underway of these and related materials and the phenomena responsible.

Conclusions

Ionic liquids are porogenic solvents for the efficient polymerization of the cross-linking monomers divinylbenzene and trimethylolpropane trimethacrylate. Little unreacted monomer is detected in the high gel-fraction ionic liquid-polymer composite by thermogravimetric analysis (made possible because of the involatility of the porogenic solvent). Solid-state NMR shows that poly(DVB) prepared in [omim]Tf$_2$N has a higher degree of cross-linked that a material prepared analogously in toluene. Significant ionic liquid retention is observed for composites formed from TRIM.

Acknowledgements

The EPSRC is thanked for financial support (through Grant Nos GR/NO5765 and GR/R15597). AIC acknowledges the support of the Royal Society through the provision of a Research Fellowship. We thank Dr. Steve Holding of RAPRA Technology for GPC data and Dr. Peter Beahan (Department of Engineering, University of Liverpool) for SEM images.

References

1. Snedden, P.; Cooper, A. I.; Scott, K.; Winterton, N. *Macromolecules* **2003**, *36*, 4549-4556.
2. Mantz, R. A.; Trulove, P. C. in Wasserscheid, P; Welton, T. Eds. *Ionic Liquids in Synthesis;* Wiley-VCH, Weinheim 2003, Section 3.2, pp56-68.
3. Vankelecom, I. F. J. *Chem. Rev.* **2002**, *102*, 3779-3810.
4. Sellin, M. F; Webb, P. B.; Cole-Hamilton, D. J. *Chem. Commun.* **2001**, 708-709.
5. Scurto, A. M.; Aki, S. N. V. K.; Brennecke, J. F. *Chem. Commun.* **2003**, 572-573.
6. Valkenberg, M. H.; deCastro, C.; Holderich, W. F. *Top. Catal,* **2001**, *14*, 139-144.
7. Huang J.; Jiang, T.; Gao, H.; Han, B.; Liu, Z.; Wu, W.; Chang, Y.; Zhao, G. *Angew. Chem. Int. Edn.* **2004**, *43*, 1397-1399.
8. Cho, T. H.; Fuller J.; Carlin, R.T. *High Temp. Material Processes* **1998**, *2*, 543-558.
9. Winterton, N, Davies, C. M.; Cooper, A. I.; Xiao, J.; Scott, K.; Basov, N. unpublished work.
10. Halpern, J. *Inorg. Chim. Acta* **1981**, *50*, 11-19.
11. Lin, H. R.; Kuo, C.-J.; Yang, C. Y.; Shaw, S.-Y.; Wu, Y.-J. *J. Biomed. Mater. Res. (Appl. Biomater.)* **2002**, *63*, 271-279.
12. Carmichael, A. J.; Haddleton, D. M. in Wasserscheid, P; Welton, T. Eds. *Ionic Liquids in Synthesis;* Wiley-VCH, Weinheim 2003, Chapter 7, pp319-335.
13. Harrisson, S.; Mackenzie, S. R.; Haddleton, D. M. *Chem. Commun.* **2002**, 2850-2851.
14. Harrisson, S.; Mackenzie, S. R.; Haddleton, D. M. *Macromolecules* **2003**, *36*, 5072-5075.
15. Kubisa, P. *Progr. Polym. Sci.* **2004**, *29*, 3-12.
16. Tiyapiboonchaiya C.; MacFarlane, D. R.; Sun, J.; Forsyth, M. *Macromol. Chem. Phys.* **2002**, *203*, 1906-1911.

17. Washiro, S.; Yoshizawa, M.; Nakajima, H.; Ohno, H. *Polymer*, **2004**, *45*, 1577-1582.
18. Benton, M. G.; Brazel, C. S. in *Ionic Liquids: Industrial Applications for Green Chemistry*, ACS Symposium Series Vol 818, Rogers, R. D.; Seddon, K. R. Eds., American Chemical Society, Washington, DC, 2002, pp125-133.
19. Sherrington, D.C. *Chem. Commun.* **1998**, 2275-2286.
20. Wasserscheid, P.; Sesing, M.; Korth, W. *Green Chem.* **2002**, *4*, 134-138.
21. Brennecke, J. F.; Anthony, J. L.; Maginn, E. J. in Wasserscheid, P; Welton, T. Eds. *Ionic Liquids in Synthesis;* Wiley-VCH, Weinheim 2003, Section 3.4, pp81-93.
22. Law, R. V.; Sherrington, D. C.; Snape, C. E. *Macromolecules* **1997**, *30*, 2868-2875.
23. Law, R. V.; Sherrington, D. C.; Snape, C. E. *Macromolecules* **1996**, *29*, 6284-6293.
24. Schmidt-Rohr, K.; Spiess, H. W. *Multidimensional Solid-State NMR and Polymers*, Academic Press, 1994.
25. Kolodziejski W.; Klinowski, J. *Chem. Rev.*, **2002**, *102*, 613-628.
26. Peters, E. C.; Petro, M.; Svec, F.; Fréchet, J. M. J. *Anal. Chem.* **1997**, *69*, 3646-3649.
27. Svec, F.; Fréchet, J.M.J. *Macromol. Symp.*, **1996**, *110*, 203-216.
28. Xie, S. F.; Svec, F.; Fréchet, J. M. J. *J. Polym. Sci., Pol. Chem.* **1997**, *35*, 1013-1021.

Chapter 10

Synthesis, Characterization, and Application of Cross-Linked Poly(ethylene glycol) Networks Used for the Gelation of Ionic Liquids

Marc A. Klingshirn, Scott K. Spear, John D. Holbrey, Jonathan G. Huddleston, and Robin D. Rogers*

Center for Green Manufacturing and Department of Chemistry, The University of Alabama, Tuscaloosa, AL 35487

Cross-linking polymerization of functionalized poly(ethylene glycol) (PEG) monomers has been found to effectively gel hydrophobic ionic liquids (ILs). By coupling a tetra-arm PEG amine with a bifunctional succinimidyl PEG monomer, stable, free-standing gels can be formed. Gelation showed a strong IL anion dependence, whereas changing the IL cations gave little change. The gels showed a shrink-swell response to their solvent environment, while the IL entrained within the gels retains bulk properties of the neat IL. Applications include potential separations media and as IL-polymer electrolytes.

Background

Poly(ethylene glycol) (PEG) is a verstatile polymer that can be used in chemical synthesis as a solvent, solvent additive, and also as a component for the formulation of various types of materials (*1-3*). PEG hydrogels have seen wide use in medical applications such as wound dressings and artificial tissue matrices (*4,5*) Their general responsiveness to solvent environments, via shrinking and swelling, also make them useful as drug delivery agents (*6*). Our group has recently reported the aqueous responsiveness of a PEG hydrogel synthesized from functionalized PEG monomers and its comparison to and applications in the area of aqueous biphasic systems (ABS) (*7*). Traditionally, these systems are comprised of an aqueous PEG phase and an aqueous salt phase. By choosing appropriate concentrations, the two phases become immiscible resulting in a biphase (*8*).

Ionic liquids (ILs) are currently being considered for replacements to volatile organic compounds in part, due to their negligible vapor pressure (*9,10*). The toxicity of the IL can also be minimized by correct choice of the cation and anion pair (*11,12*). Ionic liquids have seen use as an organic phase substitute in solvent extraction and as an electrolyte in various electrochemical applications. Non-volatility, large electrochemical windows and high thermal stability make them ideal candidates for these types of applications (*13,14*). Recent research has studied the use of ILs as a solvent for traditional polymer synthesis (free radical (*15*), living (*16*), charge transfer (*17*), palladium catalyzed (*18*)) and for the synthesis of polymers for polymer gel electrolytes (*19,20*).

The gelation of solvents, not only ILs, can be achieved by a number of different routes: by polymerization, by inter- or intramolecular forces between gelling molecules, or by reaction of functionalized monomers to give the gelled materials (*21,22*). Traditional synthesis of polymers in ILs for polymer gel electrolytes has been by *in situ* radical polymerization of vinyl monomers (*23,24*). In contrast, we have investigated the use of functionalized PEG monomers and the nucleophilic attack of amine functionalities on succinimidyl moieties that exist on the end groups on the PEG chains (*25*). This reaction enables cross-linking of PEG monomers, in turn causing gelation of the IL. In these reactions, the nature of the IL plays a significant role in the gelation process, and can, in some instances prohibit gelation all together. In addition, when exposed to aqueous environments, the relative hydrophobicity of the IL controls the degree and rate of leaching from the polymeric network. Gelled ILs have initially been studied as a new solid phase electrolyte and as a new extraction medium for organic alcohols.

Here we report the formation of IL-PEG gels with a range of ILs using the gelling system 4-arm PEG amine and disuccinimidyl PEG. The physical properties of the gels formed were investigated in comparison to control PEG-hydrogels and ILs with regard to potential applications as organic solvent separation systems and IL-gel electrolytes.

Experimental

Functionalized PEG monomers were purchased and used as received from Nektar (formerly Shearwater Polymers, Huntsville, AL). Lithium bis(trifluoromethanesulfonyl)imide salt was provided by 3M Corporation (Minneapolis, MN), while all other chemicals were purchased from Aldrich (Milwaukee, WI) and used as received. The ^{14}C-labeled linear alcohols (C_1-C_5), used in partitioning studies, were obtained from Sigma (St. Louis, MO). All water was deionized using a Barnstead Nanopure system (Dubuque, IA).

Trihexyltetradecylphosphonium chloride was provided by Cytec Inc. (Houston, TX). The IL 1-hexyl-3-methylimidazolium bis(trifluoromethanesulfonyl)imide was prepared by the method of Huddleston et al, (26), while the N-hexylpyridinium bis(trifluoromethanesulfonyl)imide IL was prepared in a similar manner with N-hexylpyridinium bromide used as starting material. 1-butyl-3-methylimidazolium tetrafluoroborate, 1-hexyl-3-methylimidazolium chloride, and 1-butyl-3-methylimidazolium hexafluorophosphate (and its hexyl analog) were prepared following methods of Bonhôte et al. and Holbrey and co-workers (14,27). Finally, 1-butyl-3-methylimidazolium trifluoromethylsulfonate was prepared in the same manner as the IL, 1-ethyl-3-methyl-imidazolium trifluoromethylsulfonate, which was previously described by Bonhôte and co-workers (14).

All hydrophobic bis(trifluoromethanesulfonyl)imide ILs were washed with copious amounts of water until the aqueous phase showed no evidence of silver halide when treated with aqueous 0.01 M silver nitrate. The resulting purity of the IL is sufficient for this work, while conventional elemental analysis would not afford reliable numbers due to the trace amounts of halide. Water contents were determined using Karl Fisher titration and were less than 0.3 M. ^{1}H and ^{13}C NMR, in addition to differential scanning calorimetry (DSC) and thermogravimetric analysis (TGA), were used to determine purity of the ILs.

Table 1 shows the amounts of PEG monomers, ILs, buffer, and IL water content used for the synthesis of the IL-PEG gels and control PEG gels. *For all studies, unless otherwise stated, the composition of IL gel 2 was used.*

DSC experiments (TA Instruments DSC 2920 Modulated DSC calorimeter, New Castle, DE) were completed using aluminum pans (TA Instruments pt. # 900793.901) with pin-hole lids (TA Instruments pt. # 900860.901). The experimental protocol followed involved initially cooling the sample to −140 °C at 5 °C min[-1], followed by heating to 400 °C with a 5 °C min[-1] heating ramp. TGA experiments (TA Instruments 2950 Thermogravimetric Analyzer, New Castle, DE), were performed under nitrogen atmosphere with heating from room temperature to 600 °C at 5 °C min[-1].

Table 1. Gel compositions used in this study

Gel	PEG 1[a] (mg)	PEG 2[a] (mg)	IL (mL)[b]	Buffer (mL)[b]	IL Water Content (mg)[c]
IL Gel 1	25.0	37.5	1.2	-----	3.5
IL Gel 2	50.0	75.0	1.2	-----	3.5
IL Gel 3	100.0	150.0	1.2	-----	3.5
Control	50.0	75.0	-----	1.2	-----

[a]PEG 1 = di-succinimidylpropyl PEG (~70 ethylene oxide units); PEG 2 = 4-arm tetra amine PEG (~225 ethyelene oxide units). [b]Each monomer was dissolved in 0.6 mL of IL or phosphate buffer independently, then combined and vortexed; [c]The determined water content of the IL was 0.16 M.

Source: Reproduced from reference 25. Copyright 2004.

Shrink-swelling experiments were performed by placing gel samples (~1 cm^3 cubes or discs of 1 cm thickness and 1 cm diameter) in 20 mL of solvent. These were then equilibrated for 24 h on a rotating wheel. Following this, the samples were gently blotted dry and the dimensions measured and compared to the initial dimensions prior to equilibration. The volume change was then calculated.

IL conductivity measurements were made on a Fisher-Accumet AR20 conductivity meter (Atlanta, GA). AC impedance spectroscopy (CH Instruments Model 660A, Austin, TX) was used to determine the conductivity of the non-hydrated IL-PEG gels. The non-hydrated IL-PEG gel was sandwiched between two parallel stainless steel rods (area = 0.5 cm^2) which were assembled in a Plexiglas holder. Impedance measurements were made with a 5 mV excitation signal over a frequency range of 100 kHz to 0.1 Hz. The thickness of the gel was 0.2 cm. All experiments were carried out at a temperature of 293 K. Conductivity was calculated using eq. 1:

$$\sigma = \frac{l}{R \times A} \qquad (1)$$

where l is the thickness of the gel, R is the bulk resistance, and A is the cross-sectional area of the electrode.

Alcohol partitioning was carried out by pre-equilibrating (rotating wheel) a known mass of non-hydrated IL-PEG gel with 5 mL of salt solution. The gel was then magnetically stirred in 1 mL of fresh salt solution and spiked with 1-4 µCi of ^{14}C-labelled tracer. The distribution coefficient was determined by determining the activity of the aqueous salt phase prior to equilibration and again after contact with the gel using eq. 2:

$$D_w = \frac{A_o - A_f}{A_f} \quad \frac{contact\ volume\ (mL)}{mass\ of\ gel\ (g)\ x\ dwcf} \tag{2}$$

where A_o is the initial activity, A_f is the activity after equilibration, and dwcf is the dry weight conversion factor, which is equal to one since the IL-PEG gel used is non-hydrated.

Results and Discussion

The IL-PEG gels, as initially synthesized are free-standing, rubbery, elastic materials that are visually transparent. However, upon submersion of the gels containing hydrophobic ILs in water, they become opaque, white in appearance due to phase separation between the IL and water. Immiscibility of the ILs and water leads to formation of hydrophobic and hydrophilic domains within the cross-linked PEG matrix.

Overall the bulk properties of the IL do not appear to be affected by the incorporation of the cross-linked PEG network and the gelation process. Figure 1 shows the DSC traces of neat 1-hexyl-3-methylimidazolium bis(trifluoromethanesulfonyl)imide ([C$_6$mim][[NTf$_2$]) and the gelled IL. After cooling and rewarming, the glass transition (1) at −85 °C, crystallization (2) at −36 °C and melting point (3) at −10 °C can all be observed in both the neat and gelled IL. The presence of these thermal events in the gelled IL indicate that the ions have little interaction with the PEG network and are essentially free to migrate. The DSC trace for a hydrated IL-PEG gel shows the freezing and melting of water within the gel. The freezing of the ionic liquid is suppressed in the system by contacting with water, in the same way that it is in water-equilibrated bulk hydrophobic ILs.

Thermal stability of the gels was assessed by thermogravimetric analysis. As synthesized IL-PEG gels show stability to above 350 °C, followed by

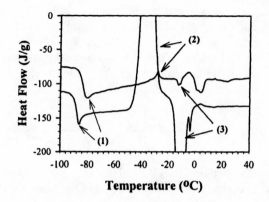

Figure 1. DSC thermograms of [C₆mim][NTf₂] PEG gel (top trace) and neat [C₆mim][NTf₂] (bottom trace).

decomposition, initially of the PEG network, followed by the IL. Gels that had been hydrated by contacting with water also showed thermal loss of absorbed water. The TGA traces show the loss of unbound water (~50 °C), bound water (~100 °C), and thermal decomposition of the cross-linked PEG matrix (~350 °C) and IL (~450 °C). Figure 2 shows the derivative weight loss curves for the as-synthesized IL-PEG gels along with their hydrated form, the neat IL, and the hydrated control PEG gel.

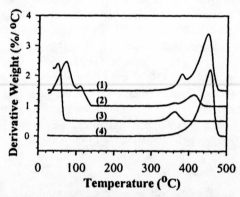

Figure 2. Derivative weight loss curves of [C₆mim][Tf₂N]-PEG gels: (1) Non-hydrated [C₆mim][Tf₂N]-PEG gel, (2) Hydrated [C₆mim][Tf₂N]-PEG gel, (3) Hydrated PEG gel control, and (4) neat [C₆mim][Tf₂N]. Curves have been offset by 0.5 %/°C for clarity.

Shrink-swell response behavior of the IL-PEG gels follows closely the behavior of the parent PEG gel synthesized in phosphate buffer. Upon equilibration of the gels in organic solvents and inorganic salts, the gels respond according to the nature of the solvent whether it is highly solvating of the PEG chains or highly dehydrating of the gel matrix. Figure 3 shows the behavior of hydrated and as-synthesized control PEG gel and the IL-PEG gel. The percent volume change is reported as compared to the initial volume of each gel prior to equilibration. The IL-PEG gel, compared to the parent PEG gel shows a similar response.

In organic solvents such as methylene chloride and methanol, swelling of the gels is generally observed. This is believed to mirror the co-miscibility seen in IL/ethanol/water systems which we have previously reported (28,29). In contrast, in the presence of high concentrations of $(NH_4)_2SO_4$ and K_3PO_4, the gels shrink due to the dehydrating nature of the salt. These observations are analogous to the salting-out effect seen in aqueous biphasic systems, since these salts are known to form biphasic systems with PEG and with ILs at high concentrations, and are highly water structuring (30,31).

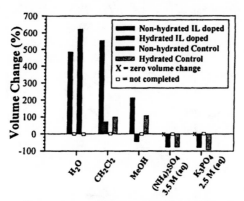

Figure 3. Shrink-swell responses of [C_6mim][NTf_2]-PEG gel and control PEG gel in various salt and organic solvents.

To help understand the effects of salt type on the shrink-swell response of the gels, a series of studies were conducted using varying concentrations of chaotropic and kosmotropic salts, which respectively, do, and do not induce biphase formation with PEG solutions. Figures 4 and 5 show the effect of anion on the volume change of the gels. The anion is known to play the major role in determining the salting-out ability of a salt, with the Gibb's free energy of hydration (ΔG_{hyd}) being a good predictor for this occurrence. Water structuring salts such as K_3PO_4 (ΔG_{hyd} for $[PO_4]^{3-}$ = -2765 kJ/mol) are better salting-out agents than chloride salts (ΔG_{hyd} for Cl^- = -340 kJ/mol) (32,33). The trends in

156

Figure 4 generally show that as the concentration of salt is increased, a greater decrease in volume change is observed. In addition, the trend of volume decrease follows that of the anion's ΔG_{hyd}. These trends are consistent with the control PEG gel under the same conditions (Figure 5).

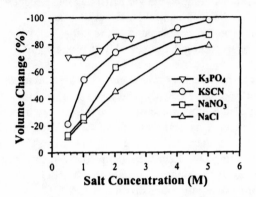

Figure 4. Response of [C₆mim][NTf₂]-PEG gel in the presence of varying salt types.

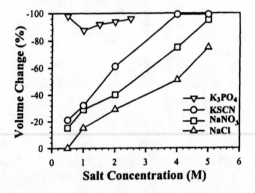

Figure 5. Response of control PEG hydrogel in the presence of varying salt types.

The data shown in Figures 4 and 5 reveals that when the IL is incorporated into the gel matrix, the decrease in volume is greater than when the gel is synthesized in the traditional phosphate buffer. This gives some clues into the solvation environment of the PEG chains when both the IL and water are present in the same matrix. Since a lesser degree of volume change is observed in the control PEG gel, it can be assumed that the removal of water from the matrix reaches a maximum, leaving only tightly bound water within the residual matrix.

When the gel is initially synthesized in the IL, hydrated in water, and then subjected to a high salt environment, dehydration of the gel matrix occurs to a much greater extent. We believe this to be due to the IL serving as the preferred solvating medium within the matrix. Since the IL is hydrophobic in nature, preferred exclusion of water from the IL, and thus the gel, is favorable, thus allowing for more water to be removed from the matrix and thus, a larger degree of negative volume change.

Attempts to form free-standing gels with other ionic liquids and solvents produced different results. Depending on the IL and solvent studied (shown in Figure 6) either pastes, readily leachable gels, or 'stable' gels were formed. The term 'stable' gel indicates that the IL diffuses out of the gel matrix at a slow rate. In some instances, however, no gel formation occurred and the fluidity of the IL is maintained. Comparing these results with the types of ILs in each case seems to indicate that hydrophobicity is a crucial element in determining whether a free-standing gel forms. In addition, the coordinating ability of the anion also seems to play a role, with the less coordinating anions ($[NTf_2]^-$ and $[PF_6]^-$) assisting gelation. The IL cation seems to play little role in determining whether an IL is capable of being gelled (both the imidazolium and pyridinium based cations with the $[NTf_2]^-$ anion are easily gelled; however, this does not transfer directly to the metal salts).

When the sodium and lithium salts of the $[NTf_2]^-$ and $[TfO]^-$ anions are tested, jelly-like masses with no structural rigidity are formed. This indicates that the cation of the metal salt may be the deciding factor in the gelation process. It is known that short chained PEGs will coordinate with lithium and sodium by wrapping around the cation to form a cavity-like structure (*34,35*). While the PEG chains in the functionalized monomers are much longer, the same type of coordination may be taking place, which essentially blocks the PEG monomers from reacting with one another by blocking the reactive end groups. Interestingly, though, gelation of the organic solvents benzene and acetonitrile takes place readily forming characteristic clear, rubbery gels. The organic solvent can, however, be removed from the matrix by evaporation, leaving opaque, white, discs.

The conductive nature of ILs makes them attractive for study as electrolytes and in solid impregnated ionic materials (*36,37*). Many types of polymeric IL electrolytes have been synthesized primarily using vinyl type monomers which proceed through free radical polymerization. In our studies, the monomers which have been incorporated follow a nucleophilic attack of the pendant amines on the succinimidyl groups to create the cross-linked networks. The conductivity of the resulting gels which incorporate $[C_6mim][NTf_2]$ show only a slight drop in conductivity, compared to the neat IL.

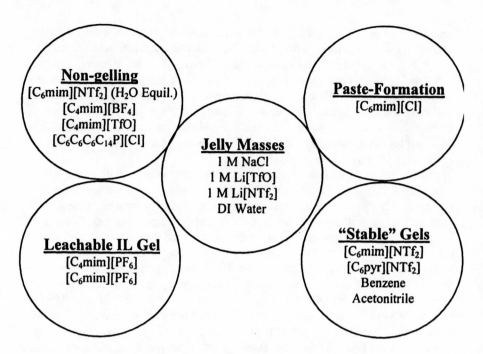

Figure 6. Attempted gelation of various solvents

Figure 7 shows the relationship between conductivity of the IL-PEG gels and increasing amounts of PEG polymer matrix between different gels. While a decrease in conductivity is initially seen when the PEG matrix is introduced, the difference between this and the neat IL is relatively small. It is interesting to note; however, that if the concentration of PEG is increased further, a critical point is seen in which no further drop in conductivity occurs.

The use of ILs as organic phase replacements in solvent extraction has been studied in depth by our group and others (*38-41*). The aqueous control PEG gel, was previously shown to display increased uptake of the organic colorant, amaranth, in response to an increase in temperature or an increase in salt concentration. The polarity of this gel phase was characterized using Reichardt's betaine dye (*7*).

PEG-salt ABS, for which the aqueous PEG-gel may be a solid phase analog, have been applied widely in protein separations and more recently in small organic and ionic separations technologies (*42,43*). The response of solvatochromatic dyes, the application of Linear Solvent Energy Relationships (LSER), and the determination of free energy of transfer of a methylene group, the latter based on the distribution of linear alcohols (*44*), have been reported.

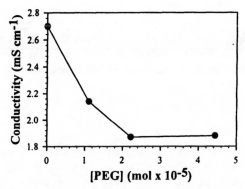

Figure 7. Conductivity of [C₆mim][NTf₂]-PEG gels at 20 °C with increasing amounts of cross-linked PEG gel matrix.

(Reproduced from reference 25. Copyright 2004.)

In a similar manner, the distribution of a series of linear alcohols from aqueous K_3PO_4 solution to the IL-PEG gels was investigated. Figure 8 presents data on the partitioning of methanol, ethanol, propanol, butanol, and pentanol to an IL-PEG gel containing [C₆mim][NTf₂]. The more hydrophobic, long, alkyl chains partition more favorably to the IL-PEG gel phase demonstrating that the gel-phase is more hydrophobic than the salt solution.

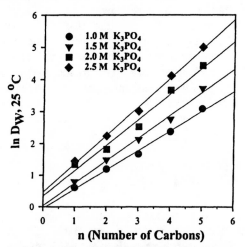

Figure 8. Natural logarithm (ln) of the weight distribution ratio of linear aliphatic alcohols ($C_nH_{2n+1}OH$) to the [C₆mim][NTf₂]-PEG gel from various K_3PO_4 solutions. The IL-PEG gel was prepared following gel formulation 2 described in Table 1.

Distribution of the alcohols increases with chain length, and with increase in salt concentration. This behavior mimics that found in conventional ABS and indicates that the IL-PEG gel/salt system can be tailored to meet specific separations needs merely by changing the salt concentration.

Conclusions

Ionic liquids can be gelled using a cross-linked PEG network formed from the reaction of functionalized PEG monomers. The resulting gels are free-standing and rubbery in nature. The ILs capable of being gelled appear to show an anion dependency with less-coordinating anions allowing for gelation. Sodium and lithium salts of the $[NTf_2]^-$ anion form jelly-like materials. The lack of complete gelation is believed to be due to cation-PEG complexation, preventing the reactive end groups from forming a complete cross-linked network. The gelled ILs, as evident from differential scanning calorimetry exhibit the same glass transition, crystallization, and melting point behavior in both gelled and bulk forms.

The uses of the gelled ionic liquids as polymer-gel electrolytes and as separations media have been studied. Their potential utility in solid electrolyte applications is promising in that the gelled materials only show a small decrease in conductivity which is consistent with the incorporation of the cross-linked matrix and its interruption of electrical flow.

Partitioning of linear alcohols to the IL-PEG gel phase from K_3PO_4 solutions indicate that the gel phase is more hydrophobic in nature than the salt solutions. This is evident from the increased partitioning of longer chain alcohols vs. their short-chain analogs. The greater discrimination between the partitioning of alcohols as the concentration of salt is increased, gives reason to believe that the system can be tailored to meet a specific separation need, simply by changing the salt concentration. This and other applications are currently being pursued.

Acknowledgments

This research was supported by the Division of Chemical Sciences, Geosciences, and Biosciences, Office of Basic Energy Research, U.S. Department of Energy (Grant DE-FG02-96ER14673) and the U.S. Environmental Protection Agency's STAR program through grant number RD-83143201-0. (Although the research described in this article has been funded in part by EPA, it has not been subjected to the Agency's required peer and policy review and therefore does not necessarily reflect the views of the Agency and no official endorsement should be inferred.)

References

1. Namboodiri, V. V.; Varma, R. S. *Green Chem.* **2001**, *3*, 146.
2. Komoto, I.; Kobayashi, S. *Chem. Commun.* **2001**, 1842.
3. Jeong, B.; Wang, L.-Q.; Gutowska, A. *Chem. Commun.* **2001**, 1516.
4. Park, K. R.; Nho, Y. C. *Rad. Phys. Chem.* **2003**, *67*, 361.
5. Hoffman, A. S. *Adv. Drug Deliv. Rev.* **2002**, *54*, 3.
6. Kissel, T.; Li, Y.; Unger, F. *Adv. Drug Deliv. Rev.* **2002**, *54*, 99.
7. Huddleston, J. G.; Looney, T. K.; Broker, G. A.; Griffin, S. T.; Spear, S. K.; Rogers, R. D. *Ind. Eng. Chem. Res.* **2003**, *42*, 6088.
8. Rogers, R. D.; Eiteman, M. A. *Aqueous Biphasic Separations: Biomolecules to Metal Ions;* Plenum Press: New York, 1995.
9. *Ionic Liquids in Synthesis*; Wassercheid. P., Welton. T., Eds.; VCH-Wiley: Weinheim, 2002.
10. *Ionic Liquids; Industrial Applications for Green Chemistry;* Rogers, R. D., Seddon. K. R.; Eds.; ACS Symposium Series 818; American Chemical Society: Washington DC, 2002.
11. Swatloski, R. P.; Holbrey, J. D.; Rogers, R. D. *Green Chem.* **2003**, *5*, 361.
12. Swatloski, R. P.; Holbrey, J. D.; Memon. S. B.; Caldwell, G. A.; Caldwell, K. A.; Rogers, R. D. *Chem. Commun.* **2004**, 668.
13. Visser, A. E.; Swatloski, R. P.; Reichert, W. M.; Mayton, R.; Sheff, S.; Wierzbicki, A.; Davis, Jr., J. H.; Rogers, R. D. *Chem. Commun.* **2001**, 135.
14. Bonhôte, P.; Dias, A.-P.; Papageorgiou, N.; Kalyanasundaram, K.; Grätzel, M. *Inorg. Chem.* **1996**, *35*, 1168.
15. Zhang, H.; Hong, K.; Mays, J. *Macromolecules.* **2002**, *35*, 5738.
16. Carmichael, A. J.; Haddleton, D. M.; Bon, S. A. F.; Seddon, K. R. *Chem. Commun.* **2000**, 1237.
17. Vijayaraghavan, R.; MacFarlane, D. R. *Aust. J. Chem.* **2004**, *57*, 129.
18. Klingshirn, M. A.; Broker, G. A.; Holbrey. J. D.; Shaughnessy, K. H.; Rogers, R. D. *Chem. Commun.* **2002**, 1394.
19. Fuller, J.; Breda, A. C.; Carlin R. T. *J. Electroanal. Chem.* **1998**, *459*, 29.
20. Fuller, J.; Breda, A. C.; Carlin R. T. *J. Electroanal. Chem.* **1997**, *144*, L67.
21. Estroff, L. A.; Hamilton, A. D. *Chem. Rev.* **2004**, *104*, 1201.
22. Kumar, D. K.; Jose, D. A.; Dastidar, P.; Das, A. *Chem. Mater.* **2004**, *16*, 2332.
23. Snedden, P.; Cooper, A. I.; Scott, K.; Winterton, N. *Macromolecules.* **2003**, *36*, 4549.

24. Noda, A.; Watanabe, M. *Electrochim. Acta.* **2000**, *45*, 1265.
25. Klingshirn, M. A.; Spear, S. K.; Subramanian, R.; Holbrey, J. D.; Huddleston, J. G.; Rogers, R. D. *Chem. Mater.* **2004**, *16*, 3091.
26. Huddleston, J. G.; Visser, A. E.; Reichert, W. M.; Willauer, H. D.; Broker, G. A.; Rogers, R. D. *Green Chem.* **2001**, *3*, 156.
27. Holbrey, J. D.; Seddon, K. R. *J. Chem. Soc., Dalton Trans.* **1999**, 2133.
28. Swatloski, R. P.; Visser, A. E.; Reichert, W. M.; Broker, G. A.; Farina, L. M.; Holbrey, J. D.; Rogers, R. D. *Chem. Commun.* **2001**, 2071.
29. Swatloski, R. P.; Visser, A. E.; Reichert, W. M.; Broker, G. A.; Farina, L. M.; Holbrey, J. D.; Rogers, R. D. *Green Chem.* **2002**, *4*, 81.
30. Gutowski, K. E.; Broker, G. A.; Willauer, H. D.; Huddleston, J. G.; Swatloski, R. P.; Holbrey, J. D.; Rogers, R. D. *J. Am. Chem. Soc.* **2003**, *125*, 6632.
31. Huddleston, J. G.; Willauer, H. D.; Griffin, S. T.; Rogers, R. D. *Ind. Eng. Chem. Res.* **1999**, *38*, 2523.
32. Marcus, Y. *J. Chem. Soc., Faraday Trans.* **1991**, *87*, 2995.
33. Rogers, R. D.; Bond, A. H.; Bauer, C. B.; Zhang, J.; Griffin, S. T. *J. Chromatogr. B* **1996**, *680*, 221.
34. Noto, D.; Longo, D.; Muchow, V. *J. Phys. Chem. B.* **1999**, *103*, 2636.
35. Broker, G. A.; Huddleston, J. G.; Baldwin, J.; Rogers, R. D. **2004**, unpublished results.
36. Ding, J.; Zhou, D.; Spinks, G.; Wallace, G.; Forsyth, S.; Forsyth, M.; MacFarlane, D. *Chem. Mater.* **2003**, *15*, 2392.
37. Kumar, G. G.; Sampath, S. *J. Electrochem. Soc.* **2003**, *150*, A608.
38. Visser, A. E.; Swatloski, R. P. ; Rogers, R. D. *Green Chem.* **2000**, *2*, 1.
39. Holbrey, J. D.; Visser, A. E.; Spear, S. K.; Reichert, W. M.; Swatloski, R. P.; Broker, G. A.; Rogers, R. D. *Green Chem.* **2003**, *5*, 129.
40. Luo, H.; Dai, S.; Bonnesen, P. V. *Anal. Chem.* **2004**, *76*, 2773.
41. Shimojo, K.; Goto, M. *Anal. Chem.* **2004**, DOI: 10.1021/ac049549x.
42. Gehrke, S. H.; Vaid, N. R.; McBride, J. F. *Biotech. Bioeng.* **1998**, *58*, 416.
43. Willauer, H. D.; Huddleston, J. G.; Rogers, R. D. *Ind. Eng. Chem. Res.* **2002**, *41*, 1892.
44. Willauer, H. D.; Huddleston, J. G.; Rogers, R. D. *Ind. Eng. Chem. Res.* **2002**, *41*, 2591.

Chapter 11

Novel Soft Composite Materials Composed of Ionic Liquids and Single-Walled Carbon Nanotubes

Takanori Fukushima[1] and Takuzo Aida[1,2]

[1]ERATO Aida Nanospace Project, Japan Science and Technology Agency (JST), c/o National Museum of Emerging Science and Innovation, 2–41 Aomi, Koto-ku, Tokyo 135–0064, Japan
[2]Department of Chemistry and Biotechnology, School of Engineering, The University of Tokyo, 7–3–1 Hongo, Bunkyo-ku, Tokyo 113–8656, Japan

Pristine single-walled carbon nanotubes form gels when ground with imidazolium ion-based room-temperature ionic liquids. In the gels, the heavily entangled nanotube bundles are exfoliated to give much finer bundles. Phase-transition and rheological properties suggest that the gels are formed by physical crosslinking of nanotube bundles mediated by local molecular ordering of ionic liquids, rather than entanglement of nanotubes. Single-walled carbon nanotube gels of ionic liquids, thus obtained, are thermally stable and do not shrivel even under reduced pressure, because of the non-volatility of the ionic liquids, but readily undergo gel-to-solid transition on absorbent materials. The use of a polymerizable ionic liquid as the gelling medium allowed for the fabrication of a novel electroconductive polymer/nanotube composite material, which showed a significant enhancement in dynamic hardness due to a strong connectivity at the polymer/nanotube interface.

Background

Over the last decade, research areas related to single-walled carbon nanotubes (SWNTs)[1] have rapidly grown both from both fundamental and technological viewpoints, because of their extraordinary mechanical and electrical properties[2,3]. SWNTs adopt a cylindrical nanostructure with a high aspect ratio, and possess a large π-electronic surface formed by rolling-up of a two-dimensional graphene sheet. SWNTs thus formed are assembled to give bundles, which are heavily entangled forming three-dimensional networks. These structural features certainly provide SWNTs with unique characteristics, but they also make SWNTs difficult to process, which hampers the huge potential of these tubular carbon nanoclusters. Adsorption of organic molecules on SWNTs via van der Waals and π-stacking interactions has been investigated in order to modify their chemical and physical properties and to improve their processability[4,5]. However, the development of much more facile and practical processing methods still remains a challenging issue.

Room-temperature ionic liquids (ILs) have attracted great attention as green recyclable alternatives to conventional organic solvents for wet processes such as chemical syntheses[6], catalyses[7], liquid/liquid extractions[8], and so forth[9-12]. We were motivated to explore the possibility of processing SWNTs with ILs, because organic cations potentially interact with π-electronic compounds through the so-called cation-π interaction[13]. Furthermore, ILs are fluid at room temperature and can be directly utilized as media for processing. Here we report the interesting phenomenon that ILs of imidazolium ions, upon grinding with SWNTs, form physical gels, where a local molecular ordering of imidazolium ions around the π-electronic SWNT surface triggers some physical crosslinking of exfoliated nanotube bundles[14]. Upon lowering the temperature, such a molecular ordering propagates further to form a crystalline material consisting of a long-range, unimodal orientation of IL molecules.

Experimental

Materials. High-purity (> 95%) HiPco SWNTs were obtained from Carbon Nanotechnologies Inc. and used throughout this work unless otherwise noted. 1-Ethyl-3-methylimidazolium tetrafluoroborate ([C$_2$mim][BF$_4$]) was purchased from Aldrich and used as received. Other ILs such as 1-butyl-3-methylimidazolium tetrafluoroborate ([C$_4$mim][BF$_4$])[15], 1-hexyl-3-methylimidazolium tetrafluoroborate ([C$_6$mim][BF$_4$])[15], 1-butyl-3-

methylimidazolium hexafluorophosphate ([C$_4$mim][PF$_6$])[16], 1-Ethyl-3-methylimidazolium bis(trifluoromethylsulfonyl)imide ([C$_2$mim][Tf$_2$N])[17], and 1-butyl-3-methylimidazolium bis(trifluoromethylsulfonyl)imide ([C$_4$mim][Tf$_2$N])[17] were prepared according to literature methods, and unambiguously characterized. A polymerizable ionic liquid, 1-(4-acryloyloxybutyl)-3-methylimidazolium hexafluorophosphate ([abmim][PF$_6$]), was prepared by alkylation of 1-methylimidazole with 4-bromobutyl acrylate followed by anion exchange of bromide by hexafluorophosphate[14].

Measurements. Electronic absorption spectra were recorded on a JASCO model V–570 UV/VIS/NIR spectrophotometer. Transmission electron micrographs (TEM) were recorded on a Philips model Tecnai F20 electron microscope operating at 120 kV. Differential scanning calorimetry (DSC) was performed on a Mettler–Toledo model DSC 822^e differential scanning calorimeter, where temperature and enthalpy were calibrated with In (430 K, 3.3 J mol^{-1}) and Zn (692.7 K, 12 J mol^{-1}) standard samples using sealed Al sample pans. Cooling and heating profiles were recorded and analyzed using the Mettler–Toledo STAR^e software system. X-ray diffraction (XRD) was performed on a Cu substrate at a scan rate of 2° min^{-1} using a Rigaku model RINT–Ultima II system equipped with a temperature controller. Dynamic rheological measurements were performed on a Rheometric Scientific model ARES with a cone-plate geometry (diameter, 2.5 cm; cone angle, 0.04 rad).

Dynamic Hardness Measurement. Sheet samples (0.6-mm thick and 1-cm^2 wide) of an [abmim][PF$_6$] polymer/SWNT (3.8 wt %) composite and an [abmim][PF$_6$] polymer were prepared by compression molding upon heating at 200 °C for 2 min with a Teflon moulder, followed by cutting into small pieces, which were tested using a Shimadzu model DUH-201S Dynamic Ultra Micro Hardness Tester, with a 115° triangular pyramid indenter at 23 ± 2 °C. Dynamic hardness (DHT$_{115}$) values were calculated from test forces (P, mN) and indentation depths (h, μm) by using the equation DHT$_{115}$ = 3.8584 P/h^2. The DHT$_{115}$ values were measured at three different points on each sample and averaged to give 0.074 and 0.019 for the composite and [abmim][PF$_6$] polymer, respectively.

Electrical Conductivity Measurement. The electrical resistance of an [abmim][PF$_6$] polymer/SWNT composite was measured by a standard four–probe method on a sheet sample. Gold wires (20 □m diameter) were attached to the sheet with a gold paste. A direct current was generated by an Advantest model R6142 power source, and the potential difference was measured on an Advantest model R6452A digital multimeter.

Results & Discussion

Gelation of ILs with SWNTs

SWNT gels of ILs, which may be called "bucky gels of ionic liquids", can be readily prepared by grinding suspensions of SWNTs in ILs, followed by removal of the excess liquid component. Typically, 2 mg of high-purity HiPco were suspended in 0.6 mL of imidazolium ion-based ILs, such as [C$_4$min][BF$_4$], (Fig. 1a), and the mixture was ground with an agate mortar for 15 minutes. The initial suspension gradually turns into a glossy black paste, which was subjected to centrifugation at 9,100g for 3 hours. A transparent liquid phase, identified as pure [C$_4$min][BF$_4$], separated from a black, lower phase containing both [C$_4$min][BF$_4$] and SWNTs. The black phase did not flow downwards when the system, after centrifugation, was inverted because of its enhanced viscosity (Fig. 1b). When the amount of [C$_4$min][BF$_4$] was varied between 0.2–1.0 mL (240–1220 mg) under otherwise identical conditions to that described above, the colorless upper phase, which separated after centrifugation, was increased in volume, while the quantity of the black, lower phase remained almost unchanged (240 mg). Furthermore, when the amount of [C$_4$min][BF$_4$] in the starting suspension was less than 0.2 mL, no ionic liquid phase separated even upon prolonged centrifugation after grinding. These observations clearly indicate that SWNT gelates the ionic liquid [C$_4$min][BF$_4$], where the gel phase is estimated to entrap up to 3×10^{20} molten salt molecules per 1 mg of SWNTs (1 wt %).

Bucky gels of ILs, thus obtained, can be readily processed. For example, one can fabricate a bucky gel string by extrusion of the gel from a syringe, and this string is not easily broken even when suspended (Fig. 1c). In sharp contrast with ordinary organo- and hydro-gels, bucky gels are highly stable and can retain their physical properties even under reduced pressure because of the non-volatility of ionic liquids. However, when placed on a filter paper, they immediately lose the entrapped ionic liquids by absorption, and turn into black powdery solids. When compared with grinding, sonication was ineffective for the gelation, and no phase transition was observed for a suspension of SWNTs in [C$_4$min][BF$_4$] after sonication (180 W) for 2 h at 30 °C.

Bucky gels were also obtained from other ionic liquids such as [C$_2$min][BF$_4$], [C$_6$min][BF$_4$], [C$_4$min][Tf$_2$N], or [C$_4$min][PF$_6$] (Fig. 1a). The former two molten salts contain the BF$_4^-$ ion, similarly to [C$_4$min][BF$_4$], and required 1 wt % of SWNTs for gelation, while the latter two salts with a [C$_4$min]$^+$ ion gelled more efficiently with 0.5 wt % of SWNTs to give swollen gels. We also found that the purity of SWNTs affects the gelation. A raw-production HiPco SWNTs, contaminated with a considerable amount (20 wt %)

of granular metal catalyst residues, brought about gelation much less efficiently, where [C₄min][BF₄] gelled with 2.5 wt % of crude SWNTs. Similarly, [C₄min][BF₄] formed a gel upon grinding with 1.5 wt % of laser-ablation SWNTs contaminated with 30 wt % of graphite. These results are summarized in Table 1.

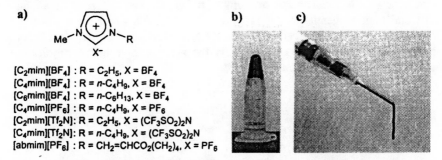

a)

[C₂mim][BF₄] : R = C₂H₅, X = BF₄
[C₄mim][BF₄] : R = n-C₄H₉, X = BF₄
[C₆mim][BF₄] : R = n-C₆H₁₃, X = BF₄
[C₄mim][PF₆] : R = n-C₄H₉, X = PF₆
[C₂mim][Tf₂N]: R = C₂H₅, X = (CF₃SO₂)₂N
[C₄mim][Tf₂N]: R = n-C₄H₉, X = (CF₃SO₂)₂N
[abmim][PF₆] : R = CH₂=CHCO₂(CH₂)₄, X = PF₆

b)   c)

Figure 1. a) Molecular formulas and schematic structures of ILs. b) Phase-separation behavior of the bucky gel of [C₄min][BF₄]. c) Extrusion of the gel from a syringe (1.0 mL).

Table 1. Critical Gelator Concentration of SWNTs

ILs	SWNTs	SWNT content (wt %)
[C₂mim][BF₄]	HiPco SWNTs (purified)[a]	0.9
[C₄mim][BF₄]	HiPco SWNTs (purified)	0.8
[C₄mim][BF₄]	HiPco SWNTs (raw)[a]	2.5
[C₄mim][BF₄]	Laser ablation SWNTs[b]	1.5
[C₆mim][BF₄]	HiPco SWNTs (purified)	0.8
[C₄mim][PF₆]	HiPco SWNTs (purified)	0.4
[C₄mim][Tf₂N]	HiPco SWNTs (purified)	0.5

[a] Purchased from Carbon Nanotechnology Inc.. Purities are estimated to be 97% and 80% for purified and raw materials, respectively. [b] This material contains ca. 30 wt % of graphite.

Characterization of Bucky Gels

SWNTs usually exist as heavily entangled bundles, which form three-dimensional networks. Upon gelation by grinding, these entangled SWNT bundles are proven to exfoliated to give much finer bundles. In TEM images, the bucky gel of [C₄min][BF₄] (Fig. 2a) showed small unentangled, bundles of SWNTs, in contrast with as-received SWNTs (Fig. 2b). Furthermore, the gel displayed well-resolved electronic absorption bands, comparable to those reported for a homogeneous thin film of SWNTs, prepared by a spray-deposition method[18]. The absorption bands of SWNTs remain unchanged upon grinding with ionic liquids. Raman spectroscopy of the gel showed the radial breathing (201 cm⁻¹) and tangential modes (1588 cm⁻¹) of SWNTs, which are essentially identical to those of as-received HiPco SWNTs (1589 and 200 cm⁻¹). From these spectral profiles, we conclude that the gelation is triggered physically without chemical denaturation of SWNTs.

a) 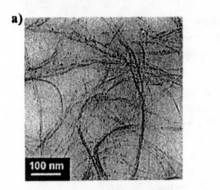 b)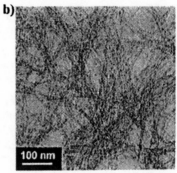

Figure 2. a) TEM image of SWNTs obtained by dispersing a gel of [C₄mim][BF₄] in deionized water. Scale bar, 100 nm. b) TEM image of as-received SWNTs.

Phase-Transision Behavior of Bucky Gels

Bucky gels of ILs show essentially different DSC and XRD profiles from those of ILs alone. The phase-transition behavior of ILs in the gel and IL alone is exemplified by the case of [C₄min][Tf₂N]. A gel of [C₄min][Tf₂N] containing 0.5 wt % of SWNTs, on cooling at a rate 10 °C/min, was supercooled reaching a glass transition point at −85 °C, while on heating, it showed sharp exothermic and endothermic peaks at −52 and −4 °C, respectively (Fig. 3a). The concentration effects of SWNTs on phase-transition behavior of the IL were also examined. As the concentration of SWNTs was increased, the exothermic peak became better defined and shifted to a lower temperature. When the system reached a critical gel (SWNTs; 0.5 wt %), the exothermic peak appeared as a single peak at −52 °C. Upon further increase in concentration of SWNTs (1.0 wt

%), the exothermic peak became much sharper without any further temperature shift. On the other hand, [C₄min][Tf₂N] alone, on heating from its glassy state, showed a very broad exothermic peak from –45 °C until the beginning of an endothermic peak at –6 °C (Fig. 3b). XRD analysis of [C₄min][Tf₂N] at –25 °C exhibited many weak diffraction peaks due to its polycrystalline structure (Fig. 3c). In sharp contrast, the mesophase between –52 and –4 °C, observed for the bucky gel of [C₄min][Tf₂N], displayed a very simple XRD pattern having an intense diffraction peak, with a d spacing of 4.60 Å (Fig. 3d).

This observation is quite interesting, considering the fact that such dramatic changes in physical properties are induced by only a 0.5 wt % of SWNTs. Since the SWNTs showed no XRD diffractions, the simplicity of the XRD profile in Fig. 3d indicates that the system consists of a unimodal, long-range molecular ordering of the molten salt without polycrystalline character. The d spacing of 4.60 Å, thus observed, is comparable with the plane-to-plane separations of interionically paired imidazolium ions, reported for single crystals of imidazolium hexafluorophosphate[19] and nitrate salts[20].

Rheological Properties of Bucky Gels

Fig. 4 shows frequency dispersion curves of dynamic storage modulus (G') and loss modulus (G'') of the gel at 25 °C. At an applied strain (γ) of 0.01, the G' curve shows a plateau region, indicating the existence of an elastic network structure in the system. Over a wide range of angular frequency (ω) from 100 to 0.04 rad/s, the G'' curve showed only a shallow minimum without any signature of relaxation, indicating that the system has permanent networks and therefore behaves as a gel. At $\gamma = 0.1$, the G' curve still shows a plateau region. However, when γ was further increased to 1.0, the system begins to behave like a critical gel, where the G' and G'' values are both reduced to low modulus levels and change with angular frequency (ω) in a power law manner without any plateau. One of the important observations in the above rheological study is the large dependence of storage modulus G' on γ at such low strain amplitudes ($\gamma < 1.0$). This feature excludes a possibility that the entanglement of SWNT bundles governs the rheological properties of the gel[21]. The systems is more likely ruled by a great number of weak physical crosslinks among the SWNT bundles, for which molecular ordering of ionic liquids is considered responsible. SWNTs can orient imidazolium ions on their π-electronic surfaces via a possible cation-π interaction[13]. Such a molecular ordering may trigger clustering of the surrounding imidazolium ions coulombically, and consequently interconnect neighboring SWNT bundles. The unimodal crystal growth in the gel at low temperature (Fig. 3d) most likely results from a controlled nucleation initiated by this local molecular orientation around the SWNT surfaces. On grinding, the large SWNT bundles are disentangled into many finer bundles, which efficiently promote the ordering of ionic liquids.

170

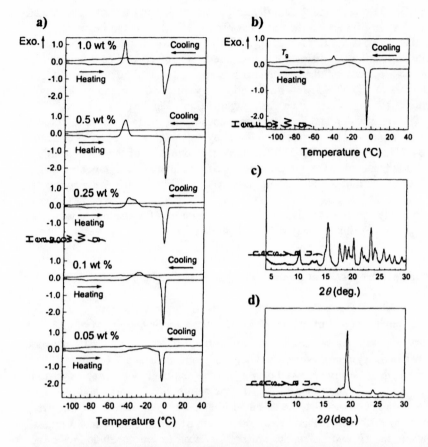

*Figure 3. DSC thermograms and XRD profiles at –25 °C on heating from –110
°C (second heating and cooling). a) mixtures of [C₄min][Tf₂N] and SWNTs
after being ground with different SWNT concentrations (wt %). b, c)
[C₄min][Tf₂N] alone. d) a bucky gel of [C₄min][Tf₂N] containing 0.5 wt % of
SWNTs. Heating and cooling rates in both DSC and XRD measurements were
10 °C min⁻¹. Prior to measurements, the samples were heated once at 100 °C
for the removal of water, if any.*

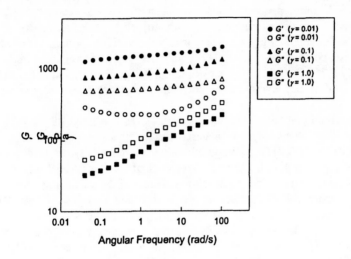

Figure 4. Angular frequency (ω) dependencies of dynamic storage (G', filled symbols) and loss moduli (G", open symbols) of a bucky gel of [C₄mim][BF₄] at 25 °C.

Fabrication and Properties of Ionic Liquid Polymer/SWNTs Composite

As an extension of the above finding, we prepared a bucky gel of a polymerizable ionic liquid and attempted *in situ* polymerization of the ionic liquid component, in order to investigate if the SWNTs included can enhance the mechanical properties of the polymer. Thus, 102 mg (3.8 wt %) of SWNTs were added to a polymerizable molten salt, [abmin][PF$_6$] (Fig. 1a; 2.56 g), and the resulting suspension, containing 2,2'-azobisisobutyronitile (AIBN, 15 mg), was ground for 1 hour to prepare a black gel. Then, the gel was heated (75 °C, 10 h), whereupon the polymerization of [abmin][PF$_6$] took place to give a homogeneous, black polymer composite of SWNTs (monomer conversion; 90%). For comparison, [abmin][PF$_6$] polymer was also prepared similarly by heating (70–75 °C, 10 h) a mixture of [abmim][PF$_6$] (1.51 g) and AIBN (10 mg) under Ar, affording a transparent polymeric substance (monomer conversion; 93%).

Interestingly, although the content of SWNTs was only 3.8 wt %, a sheet fabricated from this composite material displayed a nearly 400% increase in dynamic hardness (DHT$_{115}$) compared with a reference polymer sheet prepared solely from [abmin][PF$_6$]. The reinforcement of the bucky-gel based polymer composite is most likely due to the strong affinity of the imidazolium ion toward the π-electronic SWNT surface. We also found that the polymer/SWNT composite, thus fabricated, is electroconductive. For example, a composite sheet, containing 3.8 wt % SWNTs, showed a conductivity of 0.6 S cm^{-1}, as measured by a four-probe method. In sharp contrast, the reference polymer of [abmin][PF$_6$] without SWNTs was essentially insulating. The rather high electrical conductivity of the SWNT-reinforced polymer composite indicates its potential utility for novel electronic devices as well as coating materials, antistatic materials, and so forth.

Conclusions

We have demonstrated that room-temperature ionic liquids, when being ground with single-walled carbon nanotubes, form physical gels, whose chemical and physical properties are tunable by varying the cationic and anionic groups of the liquid components. Furthermore, this unique processing method is applicable to the fabrication of conductive polymer composite reinforced by single-walled carbon nanotubes. Considering the huge potential applications of both carbon nanotubes and ionic liquids, the present work will trigger new sciences and technologies for soft composite materials of carbon nanotubes.

Acknowledgments

The authors thank Prof. Toshikazu Takigawa of Kyoto University and Dr. Noriyuki Ishii of AIST for their active collaborations. The authors gratefully acknowledge Ms. Yayoi Taniguchi of Rigaku Corporation for low-temperature XRD measurements, and Prof. Tamotsu Inabe of Hokkaido University for conductivity measurements. The authors also thank Dr. Masako Yudasaka and Yoko Kasuya of Nanotubulites, JST-ICORP for generous supply of laser-ablation SWNTs and helpful discussions.

References

1   Iijima, S.; Ichihashi, T., *Nature* **1993**; 363, p 603.
2   Dresselhaus, M. S.; Dresselhaus, G.; Avouris, P. ed., *Carbon Nanotubes: Synthesis, Structure, Properties and Applications*; Springer–Verlag: Berlin, 2001.
3   Baughman, R. H.; Zakhidov, A. A.; de Heer, W. A., *Science* **2002**; 297, p 787.
4   Hirsch, A., *Angew. Chem. Int. Ed.* **2002**; 41, p 1853.
5   Tasis, D.; Tagmatarchis, N.; Georgakilas, V.; Prato, M., *Chem. Eur. J.* **2003**; 9, p 2.
6   Welton, T., *Chem. Rev.* **1999**; 99, p 2071.
7   Wasserscheid, P.; Keim, W., *Angew. Chem. Int. Ed.* **2000**; 39, p 3772.
8   Huddleston, J. G.; Willauer, H. D.; Swatloski, R. P.; Visser, A. E.; Rogers, R. D., *Chem. Commun.* **1998**; p 1765.
9   Freemantle, M., *Chem. Eng. News* **2001**; 79, p 21.
10 Branco, L. C.; Crespo, J. G.; Afonso, C. A.; M., *Angew. Chem. Int. Ed.* **2002**; 41, p 2771.
11 Kubo, W.; Kitamura, T.; Hanabusa, K.; Wada, Y.; Yanagida, S., *Chem. Commun.* **2002**; p 374.
12 Lu, W.; Fadeev, A. G. Qi, B.; Smela, E.; Mattes, B. R.; Ding J.; Spinks, G. M.; Mazurkiewicz, J.; Zhou, D.; Wallace, G. G.; MacFarlane, D. R.; Forsyth, S. A.; Forsyth, M., *Science* **2002**; 297, p 983.
13 Ma, J. C.; Dougherty, D. A., *Chem. Rev.* **1997**; 97, p 1303.
14 Fukushima, T.; Kosaka, A.; Ishimura, Y.; Yamamoto, T.; Takigawa, T.; Ishii, N.; Aida, T., *Science* **2003**; 300, p 2072.
15 Holbrey, J. D.; Seddon, K. R., *J. Chem. Soc., Dalton Trans.* **1999**; 2133.
16 Huddleston, J. G.; Willauer, H. D.; Swatloski, R. P.; Visser, A. E.; Rogers, R. D., *Chem. Commun.* **1998**; 1765.
17 Bonhôte, P.; Dias, A. -P.; Papageorgiou, N.; Kalyanasundaram, K.; Grätzel, M., *Inorg. Chem.* **1996**; 35, 1168.

18 Hamon, M. A.; Itkis, M. E.; Niyogi, S.; Alvaraez, T.; Kuper, C.; Menon, M.; Haddon, R. C., *J. Am. Chem. Soc.* **2001**; 123, 11292.
19 Fuller, J.; Carlin, R. T.; De Long, H. C.; Haworth, D., *J. Chem. Soc., Chem. Commun.* **1994**; 299.
20 Wilkes, J. S.; Zaworotko, M. J., *J. Chem. Soc., Chem. Commun.* **1992**; 965.
21 Doi, M.; Edwards. S. F., *The Theory of Polymer Dynamics*; Clarendon Press: Oxford, 1986.

Chapter 12

Application of Trialkylimidazolium Liquids and Salts to the Preparation of Polymer-Layered Silicate Nanocomposites and Polymer–Carbon Nanotube Nanocomposites

Douglas M. Fox[1,*], Severine Bellayer[2], Marius Murariu[2], Jeffrey W. Gilman[2], Paul H. Maupin[3], Hugh C. De Long[4], and Paul C. Trulove

[1]Chemistry Department, U.S. Naval Academy, Annapolis, MD 21402
[2]Fire Research Division, Building and Fire Research Laboratory, National Institute of Standards and Technology, Gaithersburg, MD 20899
[3]Office of Basic Energy Sciences, Office of Science, U.S. Department of Energy, Washington, DC 20585
[4]Directorate of Chemistry and Life Sciences, Air Force Office of Scientific Research, Arlington, VA 22203
*Corresponding author: telephone: 410–293–6622; fax: 410–293–2218; email: dfox@usna.edu

Polymer – layered silicate and polystyrene – carbon nanotube nanocomposites have been prepared via melt-blending. The thermal stability was analyzed using thermal gravimetric analysis and the extent of nanoparticle dispersion has been examined using powder x-ray diffraction (XRD) and transmission electron microscopy (TEM). Phase behavior of polystyrene nanocomposites were characterized using differential scanning calorimetry (DSC). The imidazolium salts used in this study were found to greatly expand the gallery d-spacing of montmorillonite to produce nanocomposites with a predominantly intercalated structure and to greatly improve the dispersion of MWNT in polystyrene to produce polymers with superior thermal properties.

Introduction

There is growing interest in improving the properties of polymers and resins through the incorporation of nanoscale materials.[1,2] These polymer nanocomposites often possess enhanced thermal stability and reduced flammability while also improving mechanical and barrier properties. In this work, we are investigating the preparation and properties of polymer – layered silicate nanocomposites (PLSNs) and polymer – carbon nanotube nanocomposites (PCNNs). To prepare PLSNs, the clay needs to be compatiblized with the polymer using an organic modifier; currently this is achieved industrially by treating the clay with dimethyl-di(hydrogenated tallow) ammonium (DMDHtAmm) salts.[2] One of the chief difficulties in the industrial production of PLSNs is the low thermal stability of DMDHtAmm salts, which limits the operating temperature of the process. The use of imidazolium compounds, specifically dimethylalkylimidazolium (DMAI) liquids and salts, in place of DMDHtAmm salts offers a significant improvement in the thermal stability of the organo-clay and allows for a higher processing temperature of the PLSNs.[2,3] Recently, we have used 1-hexadecyl-2,3-dimethylimidazolium (DMHdI) salts to produce PLSNs with the polymers poly(ethylene terephthalate) (PET) and polyamide-6 (PA-6), which require process temperatures above the onset decomposition temperatures of DMDHtAmm compounds.[2,3,4] In this work, we have prepared a new DMAI surfactant, 1,2-dimethyl-3-(benzyl ethyl iso-butyl polyhedral oligomeric silsesquioxane)imidazolium chloride (DMIPOSS-Cl) and used it to prepare organically modified layered silicates (OMS) for use in PLSNs. In addition, we have employed these DMAI surfactants to produce PCNNs. Preliminary results of the PCNNs have suggested a similar improvement in thermal stability and greater reduction in flammability over the neat polymers. We present thermal stability and dispersion results for PLSNs and PCNNs with a variety of polymers, including PS, poly(ethylene-co-vinyl acetate) (EVA) and PA-6. A summary of the abbreviations used in this paper are given in Table 1.

Experimental

RTIL Preparation

Chlorobenzyl ethyl iso-butyl-POSS (10 g, Hybrid Plastics) and a 30% excess of 1,2-dimethylimidazole (Sigma-Aldrich) were added to THF (25 ml). After dissolution, the solution was heated to a gentle reflux for 3 days. The solution was cooled, dissolved in hot ethyl acetate and precipitated with addition of acetonitrile. The solid product was washed with acetonitrile and dried under vacuum overnight. Duplicate elemental analysis performed by Galbraith Laboratories indicated the prepared DMIPOSS consisted of 46.55 %C, 7.86 %H, and 2.45 %N (theoretical = 47.3 %C, 7.7 %H, and 2.6%N). ^{1}H NMR (CDCl$_3$, 300 MHz) was characterized by the disappearance of the chloromethylene resonance from the starting material, σ = 4.5 ppm (2H,s) and the appearence of new resonances at σ = 2.7 ppm (3H,s), 3.9 pmm (3H,s), 5.4 ppm (2H, s), 7.5 ppm (1H, d) and 7.7 ppm (1H, d) corresponding to the imidazolium 2-methyl, the imidazolium 1-methyl, the imidazolium 3-benzylic methylene, and the 4 and 5 position imidazolium hydrogens, respectively. The 3.9 ppm peak was phase inverted. Assignments were made from comparision with Aldrich spectra of some simple imidazole and imidazolium compounds.

DMHdIBF$_4$ was prepared in acetonitrile, as previously described.[5] Elemental analysis by Galbraith Laboratories resulted in mass errors of 0.2% C, 1.0% H, and 1.2% N. The water content was determined to be 24 ppm using a Brinkmann Model 756/2 Karl Fischer Coulometric Titrator.

Table 1: Abbreviations for chemicals used in this study

Abbreviation	Definition
POSS	Polyhedral oligomeric silsesquioxanes
DMIPOSS	1,2-dimethyl-3-(benzyl ethyl iso-butyl polyhedral oligomeric silsesquioxane)imidazolium cation
DMHdI	1-hexadecyl-2,3-dimethylimidazolium cation
DMDOdAmm	N,N-dimethyl-N,N-dioctadecylammonium cation
MMT	montmorillonite
MWNT	Carbon multi-walled nanotubes
PS	Polystyrene
EVA	Poly(ethylene co-vinyl acetate)
PA-6	Polyamide-6
PLSN	Polymer – layered silicate nanocomposites
PCNN	Polymer – carbon nanotube naocomposites
OMS	Organically modified layered silicate

OMS Preparation

DMHdI-MMT and DMDOdAmm-MMT were prepared using standard exchange procedures as described previously.[2,3] Na-MMT was exchanged with

DMIPOSS using a slight modification to this procedure. DMIPOSS-Cl (3 g) was dissolved in THF (50 ml), then Na-MMT (3 g, Southern Clay Products) was added to form a slurry. The slurry was heated to boiling, deionized water (25 ml) was slowly added, and the mixture was removed from the heat. After stirring for 2 days, the slurry was filtered and washed with aliquots of hot THF (50 ml). The exchanged clay was leached with hot 75:25 ethanol:water, filtered, and dried at 100°C for 1 hr.

MWNT/DMHdI mixtures

MWNT/DMHdIBF$_4$ mixtures were prepared with mass fraction ratios of 1:1, 1:4, and 1:19 (MWNT to DMHdIBF$_4$). DMHdIBF$_4$ and MWNT were melt blended at 185°C in a conical co-rotation twin screw mini-extruder (DACA) at 5.2 rad/s for 2 min. The mixtures solidified upon cooling, and were ground into a fine powder for use in preparing PCNNs.

Nanocomposite Preparation

Polystyrene (PS) – carbon nanotube nanocomposites were prepared using the extruder described previously. Enough MWNT/DMHdIBF$_4$ mixture was melt blended with PS at 195°C and 26 rad/s for 5 min to prepare nanocomposites with 0.5 wt-% MWNT. Additionally, composites were prepared using 0.5 wt-% pure MWNT and 0.5 wt-% pure DMHdIBF$_4$ as PS + MWNT and PS + DMHdIBF$_4$ references. All of the nanocomposites prepared (except PS + DMHdIBF$_4$) had 0.5 wt-% MWNT, regardless of the imidazolium content. Thus, PS + (1:1) MWNT:DMHdIBF$_4$ contains 0.5 wt-% MWNT and 0.5 wt-% DMHdIBF$_4$; PS + (1:4) MWNT:DMHdIBF$_4$ contains 0.5 wt-% MWNT and 2.0 wt-% DMHdIBF$_4$; and (1:19) MWNT:DMHdIBF$_4$ contains 0.5 wt-% MWNT and 9.5 wt-% DMHdIBF$_4$.

Instrumentation

Thermal stabilities were determined using a TA Instruments, Hi-Res TGA2960 Thermogravimetric Analyzer. For the comparative TGA study, 4.0 mg to 6.0 mg samples were placed in open ceramic pans and heated at a scan rate of 10°C/min while purged with 100 ml/min N$_2$. For all TGA anlyses, the mean of three replicate measurements was typically reported. The temperature of both the onset (5% mass fraction loss) and peak mass loss rate have an uncertainty of $\sigma = \pm 4°C$. The mass loss rate has a relative uncertainty of $\sigma/\bar{x} = \pm 10\%$.

Melting and freezing properties were characterized using a TA Instruments DSC2910 Differential Scanning Calorimeter connected to a Refrigerated Cooling System. 3.0 mg to 5.0 mg samples were hermetically sealed in aluminum pans and were heated and cooled at a scan rate of 5°C/min while purged with 100 ml/min N_2. Data were collected during the second consecutive scans.

Powder x-ray diffraction experiments were performed on a Bruker AXS D8 Powder diffractometer. The d-spacing was calculated from peak positions using Cu $K\alpha$ radiation ($\lambda = 0.15418$ nm) and Bragg's Law. Clay samples were ground to a fine powder prior to loading. PLSNs samples were prepared by pressing (6 atm) 0.5 g in a 2 cm x 1.6 m mold at temperatures above the melting point. Standard x-ray measurements were performed over a scanning 2θ range of 2.5° - 12° using a step size of 0.05° for 20 sec.

All TEM samples were ultra-microtomed using a diamond knife on a Leica Ultracut UCT microtome at room temperature. The cut samples (100 nm nominal thickness) were transferred to 400 mesh Cu/Rh grids. A Philips 400T electron microscope was used to obtain bright-field TEM images at 120 kV under low-dose conditions. To ensure the collection of representative images, several images of various magnifications with 2-3 sections per grid were obtained.

Results and Discussion

Polymer – Layered Silicate Nanocomposites

The thermal stability of the organically modified clays (OMS) were determined using thermal gravimetric analysis (TGA) in nitrogen. As shown in Figure 1, DMIPOSS-MMT exhibits a thermal stability in nitrogen which is 50°C higher than 1-hexadecyl-2,3-dimethylimidazolium exchanged clay (DMHdI-MMT) and 200°C higher than the conventional N,N-dimethyl-N,N-dioctadecylammonium exchanged clay (DMDOdAmm-MMT).

To help facilitate exfoliation of the clay layers during the preparation of the PLSNs, it is desirable to exchange the Na^+ with a surfactant that will also swell the silicate layers. We used powder x-ray diffraction (XRD) to measure the expansion of the montmorillonite layers after cation exchange. The use of DMIPOSS as the organic surfactant greatly increases the d-spacing between the clay layers. As shown in Figure 2, this expansion is far greater than any previously used imidazolium treatment or the commonly used quaternary

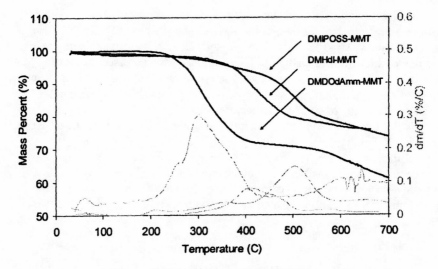

Figure 1. TGA of organically modified clays (OMC) in N_2 at 10°C/min

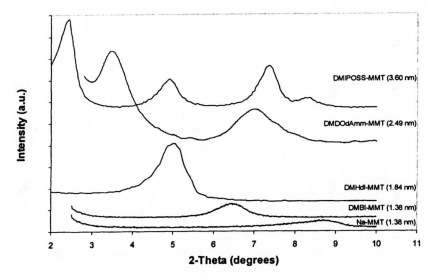

Figure 2. XRD of OMCs. Plots have been shifted upward for clarity.

ammonium treatment. It is important to note that DMIPOSS-MMT exhibits peaks at 8.3° and 10.6° (not shown) 2θ, which are characteristic POSS crystalline peaks.[6,7,8] The large d-spacing and POSS cystallinity peaks suggests that the DMIPOSS cations on adjacent clay layers may be arranged as a bilayer in the gallery with significant POSS-POSS interactions.

The thermal stabilities of the processed polymer – layered silicate nanocomposites (PLSNs) in nitrogen were measured using TGA. The thermal stabilities of polystyrene nanocomposites are shown in Figure 3. Similar to other PLSN studies, the melt blending of DMIPOSS-MMT and PS results in a significant improvement in thermal stability and an increase in char yield.[9,10] Although the use of DMIPOSS-MMT does improve the thermal characteristics of the polymers, it is not as effective as other OMS, as indicated in Table 2. The larger d-spacing of DMIPOSS-MMT, smaller expansion of the clay galleries upon blending with the polymers (Δδ), and loss of XRD peaks associated with the POSS crystallinity (not shown) all indicate that the polymers form a predominantly intercalated structure with DMIPOSS-MMT. The higher thermal stability of PLSNs using DMHdI-MMT is due to exfoliation of the clay layers (i.e. complete separation of the clay layers which is indicated by a loss of the XRD peak associated with the clay gallery spacing), which increases the amount of clay-polymer interactions.

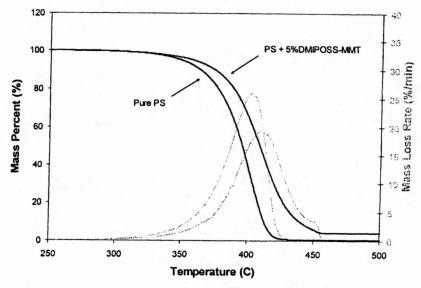

Figure 3. TGA of PS + 5%DMIPOSS-MMT in N₂ at 10°C/min

Table 2: TGA Onset and Peak Temperatures and Shifts in XRD peak associated with the gallery d-spacing of the OMCs.

Composite	T_{onset} (°C)*	T_{peak} (°C)*	$\Delta\delta$ (nm) v. O-MMT
PS (180°C, 200rpm, 5m)	346	404	---
+5% DMHdIMMT	371	428	1.7
+5% DMIPOSSMMT	360	412	0.2
EVA (170°C, 200rpm, 5m)	332	352/457	---
+5% DMHdIMMT	344	353/470	exfoliated
+5% DMIPOSSMMT	337	352/454	0.5
PA-6 (250°C, 200rpm, 5m)	383	443	---
+5% DMHdIMMT	384	438	exfoliated
+5% DMIPOSSMMT	382	435	0.2

Transmission electron microscopy (TEM) images of the PS + 5%DMIPOSS-MMT nanocomposite shown in Figure 4 reveal the presence of a significant number of multi-layer tactoids throughout the polymer. Similarly to the XRD data, the TEM images suggest significant intercalation with an average of 6-10 layers per tactoid, indicating about 15% of the layers exfoliated from the adjacent layer. The large d-spacing of the OMS and rigid structure of the POSS cages lowers the entropic gain of both the surfactant and the intercalated polymer. The lack of favorable polymer-surfactant or polymer-clay interactions plus the presence of strong POSS-POSS interactions results in a lack of favorable enthalpic conditions for exfoliation. Hence, the free energy gain for exfoliation is low, resulting in predominantly intercalated structures.[11]

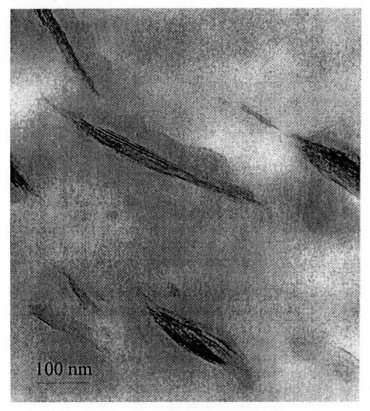

Figure 4. TEM image of PS + 5%DMIPOSS-MMT

Polymer – Carbon Nanotube Nanocomposites

The XRD of DMHdIBF$_4$ (see Figure 5) exhibits three peaks in the low angle region (2θ = 3.2°, 6.4°, and 12.8°), characteristic of the crystalline solid phase and representing the (001) reflections from interdigitated bilayer structure with an interlayer distance of 2.75 nm.[12,13] In the (1:1) MWNT:DMHdIBF$_4$ mixture, these peaks disappear and new peaks form at higher angles, representative of alkyl chain ordering in the lamellar structure.[12] The XRD of the (1:4) MWNT:DMHdIBF$_4$ and (1:19) MWNT:DMHdIBF$_4$ (not shown) mixtures exhibit peaks consistant with both structures, suggesting the mixtures consist of 2 separate phases.

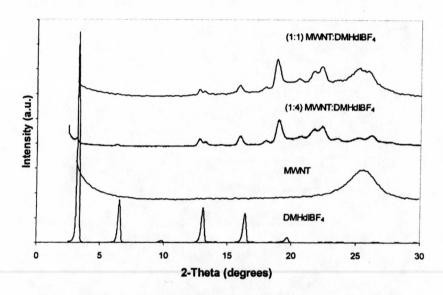

Figure 5. XRD of DMHdIBF4, MWNT, and MWNT-DMHdIBF$_4$ mixtures. Plots have been shifted upward for clarity.

The polymer nanocomposites prepared from PS and the MWNT/DMHdIBF$_4$ mixtures were characterized using DSC, TEM, and TGA. The melting characteristics were investigated using DSC and are shown in Figure 6. DMHdIBF$_4$ exhibits liquid crystalline behavior, with a melting point at 70°C and a clearing point at 175°C. Pure PS exhibits a glass transition around 100°C. The PCNN prepared from (1:1) MWNT:DMHdIBF$_4$ does not exhibit any peaks characteristic of pure DMHdIBF$_4$ and has the same glass transition temperature as the pure PS. This suggests that the DMHdIBF$_4$ is completely associated with the MWNTs. It also indicates that, similarly to the poly(methyl methacralate)/MWNT results obtained by Tatro, et. al.,[14] low loadings of MWNTs do not retard PS chain motion. PCNNs prepared using DMHdIBF$_4$ rich mixtures do have endotherms associated with DMHdIBF$_4$ melting and liquid crystal clearing points, indicating that the excess imidazolium salt has phase separated from the PS matrix.

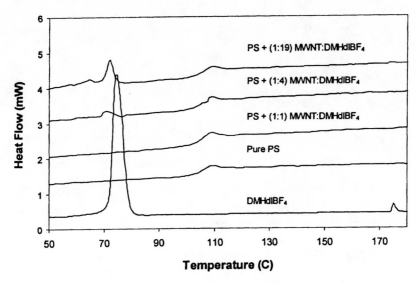

Figure 6. DSC of PS and PS-Carbon Nanotube Composites. Plots have been shifted upward for clarity.

The dispersion of MWNT and phase separation of excess DMHdIBF$_4$ can be visualized using TEM. TEM images of PS + MWNT and PS + (1:1) MWNT:DMHdIBF$_4$ are shown in Figure 7. The background was removed in a portion of the images to conduct image analyses, which indicated that the images were indeed representative of the entire samples.[12] When untreated MWNT are directly incorporated in the PS matrix, poor dispersion is obtained, with a median nearest neighbor distance of 38 ± 13 nm. The use of (1:1) MWNT:DMHdIBF$_4$ greatly improves dispersion, with the MWNT median nearest neighbor distance of 109 nm ± 6 nm. The theoretical nearest neighbor distance for a perfectly mixed sample should be 120 nm ± 15 nm for PS + untreated MWNT and 132 nm ± 20 nm for PS + (1:1) MWNT:DMHdIBF$_4$. These results illustrate the natural tendency for MWNT to agglomerate and show that DMHdIBF$_4$ interacts with both MWNT and PS to reduce this agglomeration and improve dispersion. The nanocomposites prepared using (1:4) and (1:19)

MWNT:DMHdIBF$_4$ produces MWNT dispersion similar to that of the (1:1) sample, but also reveal holes in the TEM images (not shown). This illustrates the phase separation behavior revealed in the DSC and suggests a π-stacking type interaction between the imidazolium cation and MWNT.[12]

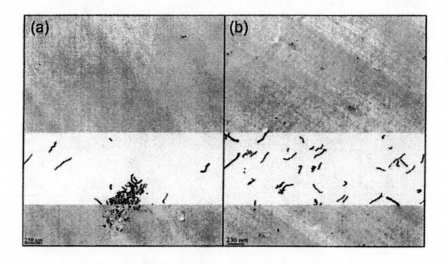

Figure 7. TEM images of (a) PS + MWNT and (b) PS + (1:1) MWNT:DMHdIBF$_4$

The thermal stability of PS + (1:1) MWNT:DMHdIBF$_4$ in N$_2$ was measured using TGA. PS, PS + MWNT, and PS + DMHdIBF4 thermal stabilities were also measured to evaluate the effectiveness of the MWNT/DMHdIBF$_4$ mixture. The thermogram is shown in Figure 8. Both the onset and maximum decomposition temperatures of PS were not affected by the addition of either MWNT or DMHdIBF$_4$. However, the addition of the (1:1) MWNT:DMHdIBF$_4$ mixture improved the thermal stability of PS by 20° ± 3°C. This stabilization may be explained by the enhanced dispersion of MWNT in this nanocomposite, which increases the contact surface between MWNT and the polymer matrix.

Conclusions

Polymer-Layered Silicate Nanocomposites
A new imidazolium compound (DMIPOSS-Cl), consisting of a POSS cage tethered to one of the nitrogen groups, was prepared and successfully used as a

surfactant for the preparation of an organically modified clay. DMIPOSS-Cl expands the galleries of the clay, montmorillonite, to a greater extent than quaternary ammonium or other imidazolium surfactants. The large d-spacing of the DMIPOSS treated clay, the strong POSS-POSS interactions within the clay galleries, and the incompressibility of the POSS cages prevent complete delamination of the clay layers in PS, EVA, and PA-6 nanocomposites, leading to predominantly intercalated structures. Thus, the thermal stability of the PS PLSN is improved, but not as much as using clays modified by other imidazolium surfactants.

<u>Polymer-Carbon Nanotube Nanocomposites</u>
Trialkylimidazolium salt compatibilized MWNTs have been used to prepare high quality PS nanocomposites using a melt-blending process. DSC, XRD, and TEM reveal that (1:1) MWNT:DMHdIBF$_4$ mixtures can be blended with PS to produce single phase nanocomposites with good MWNT dispersion. MWNT/DMHdIBF$_4$ mixtures with higher ratios of DMHdIBF$_4$ result in phase separation of the excess DMHdIBF$_4$. The PS + (1:1) MWNT:DMHdIBF$_4$ nanocomposite exhibits a thermal stability that is 20°C higher than the pure PS. Thus, well-dispersed, single phase PS nanocomposites can be prepared using a very low loading of MWNT to improve the thermal properties of the polymer.

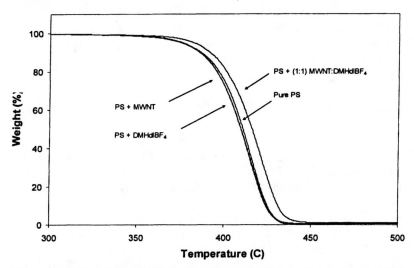

Figure 8. TGA of PS and PS-Carbon Nanotube Composites in N$_2$ at 10°C/min

References

* - The policy of the National Institute of Standards and Technology (NIST) is to use metric units of measurement in all its publications, and to provide statements of uncertainty for all original measurements. In this document however, data from organizations outside NIST are shown, which may include measurements in non-metric units or measurements without uncertainty statements. The identification of any commercial product or trade name does not imply endorsement or recommendation by NIST or the United States Air Force (USAF). Opinions, interpretations, conclusions, and recommendations are those of the authors and are not necessarily endorsed by the USAF or NIST.

† - Financial support was provided by the USAF and NIST.

1. M. Alexandre and P. Dubois, *Mater. Sci. Eng. R: Rep.*, **2000**, 28, 1.
2. W. H. Awad, J. W. Gilman, M. Nyden, R. H. Harris Jr., T. E. Sutto, J. Callahan, P. C. Trulove, H. C. De Long, and D. M. Fox, *Thermochim. Acta*, **2004**, 409, 3.
3. J. W. Gilman, W. H. Awad, R. D. Davis, J. Shields, R. H. Harris Jr., C. Davis, A. B. Morgan, T. E. Sutto, J. Callahan, P. C. Trulove, and H. C. De Long, *Chem. Mater.*, **2002**, 14, 2776.
4. C.H. Davis, L.J. Mathias, J.W. Gilman, D.A. Schiraldi, P.C. Trulove, T.E. Sutto, H.C. De Long, *J. Polym. Sci., Poly. Phys.*, **2002**, 40, 2661.
5. D. M. Fox, W. H. Awad, J. W. Gilman, P. H. Maupin, H. C. De Long, and P. C. Trulove, *Green Chem.*, **2003**, 5, 724.
6. J. B. Carroll, A. J. Waddon, H. Nakade, and V. M. Rotello, *Macromolecules*, **2003**, 36, 6289.
7. L. Zheng, A. J. Waddon, R. J. Farris, and E. B. Coughlin, *Macromolecules*, **2002**, 35, 2375.
8. A. J. Waddon and E. B. Coughlin, *Chem. Mater.*, **2003**, 15, 4555.
9. Z. M. Wang, T. C. Chung, J. W. Gilman, and E. Manias, *J. Polym. Sci. B: Polym. Phys.*, **2003**, 41, 3173.
10. S. Bourgibot, J. W. Gilman, and C. A. Wilkie, *Polym. Degrad. Stab.*, **2004**, 84, 483.
11. R. A. Vaia and E. P. Giannelis, *Macromolecules*, **1997**, 30, 8000.
12. S. Bellayer, J. W. Gilman, N. Eidelman, S. Bourbigot, X. Flambard, D. M. Fox, H. C. De Long, and P. C. Trulove, *Adv. Mater.*, submitted.
13. F. Neve, O. Francescangeli, and A. Crispini, *Inorg. Chim. Acta*, **2002**, 338, 51.
14. S. R. Tatro, L. M. Clayton, P. A. O. Muisener, A. M. Rao, and J. P. Harmon, *Polymer*, **2004**, 24, 1971.

Indexes

Author Index

Aida, Takuzo, 163
Bellayer, Severine, 175
Brazel, Christopher S., 103
Chen, Ji, 71
Cooper, Andrew I., 133
De Long, Hugh C., 175
Fox, Douglas M., 175
Fukushima, Takanori, 163
Gilman, Jeffrey W., 175
Görnitz, Eckhard, 17
Guerrero-Sanchez, Carlos, 37
Holbrey, John D., 71, 149
Hong, Kunlun, 1
Huddleston, Jonathan G., 149
Khimyak, Yaroslav Z., 133
Klingshirn, Marc A., 149
Kraudelt, Heide, 17
Laschewsky, André, 17
Ma, Hongyang, 51
Maupin, Paul H., 175
Mays, Jimmy W., 1
Murariu, Marius, 175

Noda, Akihiro, 119
Ohno, Hiroyuki, 89
Rahman, Mustafizur, 103
Rogers, Robin D., 71, 149
Schubert, Ulrich S., 37
Scott, Keith, 133
Shoff, Hugh W., 103
Snedden, Peter, 133
Spear, Scott K., 71, 149
Strehmel, Veronika, 17
Susan, Md. Abu Bin Hasan, 119
Swatloski, Richard P., 71
Trulove, Paul C., 175
Turner, Megan B., 71
Wan, Xinhua, 51
Washiro, Satoko, 89
Watanabe, Masayoshi, 119
Wetzel, Hendrik, 17
Wiesbrock, Frank, 37
Winterton, Neil, 133
Yoshizawa, Masahiro, 89
Zhang, Hongwei, 1

Subject Index

A

Acrylonitrile
compatibility of 1-ethyl-3-methylimidazolium tetrafluoroborate and 1-butylpyridinium tetrafluoroborate with, and its polymer, 125*t*
See also Polyacrylonitrile (PAN)
Alcohol partitioning
distribution of alcohols to ionic liquid-poly(ethylene glycol) gels, 159–160
procedure, 153
See also Poly(ethylene glycol) (PEG) networks
Alcohols, partitioning of linear to poly(ethylene glycol) networks, 159–160
1-Allyl-3-methylimidazolium chloride
solvent for cellulose, 74
structure, 74*f*
Atom transfer radical polymerization (ATRP)
mechanism, 52–53
See also Reverse atom transfer radical polymerization (ATRP)

B

Benzene
characterization data for polymerization of methyl methacrylate (MMA) and styrene, 6*t*
^{13}C NMR spectra of PMMA in, and ionic liquid, 8*f*
size exclusion chromatography (SEC) of PMMA , 7*f*
Benzoyl peroxide, polymerization in room temperature ionic liquids, 5, 6*t*
Bucky gels
characterization, 167–168
ionic liquids, 166–167
rheological properties, 169, 171*f*
See also Single-walled carbon nanotubes (SWNTs)
1-Butyl-3-methylimidazolium chloride
solvent for cellulose, 73–74
structure, 74*f*
structures of polymer soluble in, 82*f*
1-Butyl-3-methylimidazolium hexafluorophosphate (BMIM-PF$_6$)
atom transfer radical polymerization (ATRP) of *N*-substituted maleimides with styrene, 56
procedure for reverse ATRP of methyl methacrylate (MMA) in, 54–55
reverse ATRP of MMA at different [AIBN]/[CuCl$_2$] ratio, 58–59
reverse ATRP of MMA at different temperatures, 56, 57*t*
solubility of acrylates in, 55–56
1-Butyl-3-methylimidazolium tetrafluoroborate (BMIM-BF$_4$)
polymerization of butyl methacrylate (BMA), 42–43, 44*f*
polymerization of methyl methacrylate (MMA), 42–43, 44*f*
polymerization of styrene, 42–43, 45*f*

precipitation of PMMA, 43, 46*f*
proposal for PMMA production
using, 46, 47*f*
reverse ATRP of MMA in, 59–60
See also Water-soluble ionic
liquids (ILs)
Butyl methacrylate (BMA)
degree of polymerization in
imidazolium salts or toluene,
25*f*, 26*f*, 28*f*
experimental, 19–20
gel permeation chromatography
(GPC) of polymer, 24*f*, 26*f*, 27*f*
glass transition temperature of
polymer of, 27–28
homopolymerization and
copolymerization, 18
imidazolium salts, 20, 21*t*
kinetics of polymerization in bulk
and water-soluble ionic liquids,
42–43, 44*f*
polymerization in ionic liquid, 19
polymerization using imidazolium
salts, 24–28
polymers containing zwitterionic
structures, 29, 32
recycling ionic liquids after
polymerization, 32, 33*f*
structure, 20*f*
See also Imidazolium salts
1-Butylpyridinium tetrafluoroborate
compatibility with vinyl monomers
and their polymers, 125*t*
polymerization of ion gels, 123–
124
structure, 122
thermal properties and ionic
conductivities, 124*t*
See also Polymer electrolytes

C

Carbon nanotubes. *See* Single-walled
carbon nanotubes (SWNTs)

Cellulose
1-allyl-3-methylimidazolium
chloride as good solvent, 74
1-butyl-3-methylimidazolium
chloride [C$_4$mim]Cl to
dissolve, 73–74
cellulose/polyacrylonitrile blends,
82–83
composite and encapsulated
materials, 77–81
compositing techniques, 77
dissolution by
LiCl/dimethylacetamide solvent,
74
formation of composite polymers,
81–83
incorporation of dyes and
complexants, 77
laccase enzyme in cellulose
polymer films, 79–81
linear solvent-energy relationship
(LSER), 75–76
magnetite/cellulose composites,
78–79
molecular structure, 73*f*
scanning electron micrograph of,
fiber, 75*f*
solubility in ionic liquids, 73–
76
Chiral ionic liquids
reverse atom transfer radical
polymerization (ATRP) of
methyl methacrylate (MMA),
63–64
synthesis, 61, 63
Chloride content, imidazolium salts,
19–20, 21*t*
Composites
cellulose/polyacrylonitrile blends,
82–83
compositing techniques, 77
fabrication and properties of ionic
liquid polymer/single-walled
carbon nanotubes composite,
172

interest in ionic liquid-polymer
  composites, 134
ionic liquids for dye and
  complexant incorporation, 77
laccase enzyme in cellulose
  polymer films, 79–81
magnetite/cellulose, 78–79
See also Cellulose; Crosslinked
  polymers in ionic liquids;
  Polymer-carbon nanotube
  nanocomposites; Polymer-
  layered silicate nanocomposites
Conductivity, ionic liquid-
  poly(ethylene glycol) gels, 158,
  159*f*
Copper(II) chloride content, reverse
  atom transfer radical
  polymerization of methyl
  methacrylate, 58–59
Crosslinked polymers in ionic liquids
  characterization methods, 136–137
  composite and polymer
    characterization, 138–144
  composites of crosslinked polymer-
    ionic liquid and corresponding
    crosslinked polymers, 139*t*
  crosslinking monomers in
    polymerizations in ionic liquids,
    136*f*
  divinylbenzene (DVB)
    polymerization, 137
  experimental, 136–137
  ionic liquid-polymer composites,
    134
  monomers polymerized in ionic
    liquids, 135*f*
  poly(TRIM) characterization, 143–
    144
  polymer and composite
    preparation, 138
  polymerization of 4-vinylpyridine
    (VP) and DVB, 137
  polymerization of
    trimethylolpropane
    trimethacrylate (TRIM), 137

polymers and polymerizations in
  ionic liquids, 134–135
pore size distribution for
  poly(TRIM), 144*f*
pore size distributions for Soxhlet
  extracted poly(DVB), 140*f*
porogenic behavior, 144–145
procedures for polymerizations and
  polymer purification, 137
properties of crosslinked polymers,
  140*t*
scanning electron micrograph of
  Soxhlet-extracted poly(TRIM)s,
  144*f*
scanning electron micrographs of
  Soxhlet-extracted poly(DVB)s,
  141*f*
solid-state NMR studies of
  poly(DVB), 141–143
See also Poly(ethylene glycol)
  (PEG) networks

D

Degree of polymerization, poly(*n*-
  butyl methacrylate) using
  imidazolium salts as solvents, 25*f*,
  26*f*, 28*f*
Differential scanning calorimetry
  (DSC)
  glass transition temperatures of
    poly(methyl methacrylate)s
    (PMMAs), 9
  ionic liquid-poly(ethylene glycol)
    (PEG) networks, 151, 153, 154*f*
  polymer-carbon nanotube
    composites, 184, 185*f*
  single-walled carbon nanotubes in
    ionic liquid, 170*f*
Diffusion-controlled termination,
  polymerization mechanism, 10
Dimethylacetamide/LiCl, dissolution
  of cellulose, 74
Divinylbenzene (DVB)

polymerization procedure, 137
See also Crosslinked polymers in ionic liquids
1-Dodecyl-3-methylimidazolium tetrafluoroborate, reverse atom transfer radical polymerization of methyl methacrylate, 59–60
Dynamic storage, bucky gels, 169, 171*f*

E

Elastic moduli, plasticized poly (vinyl chloride) (PVC), 114, 115*t*
Electrolytes. *See* Polymer electrolytes; Solid polymer electrolytes
Enzymes
cellulose polymer films, 79–81
stabilization with protective pre-coating, 80–81
1-Ethyl-3-methylimidazolium bis(trifluoromethane sulfonyl)-imide (EMITFSI)
photograph of polymer electrolyte film of EMITFSI/PMMA ion gel, 126*f*
polymerization of ion gels, 123–124
structure, 122
thermal properties and ionic conductivities, 124*t*
See also Polymer electrolytes
1-Ethyl-3-methylimidazolium tetrafluoroborate
compatibility with vinyl monomers and their polymers, 125*t*
polymerization of ion gels, 123–124
structure, 122
thermal properties and ionic conductivities, 124*t*
See also Polymer electrolytes

1-Ethyl-3-methylimidazolium tosylate (EMIM-TOS)
polymerization of butyl methacrylate (BMA), 42–43, 44*f*
polymerization of methyl methacrylate (MMA), 42–43, 44*f*
polymerization of styrene, 42–43, 45*f*
precipitation of PMMA, 43, 46*f*
See also Water-soluble ionic liquids (ILs)

F

Flexibility, plasticized poly(vinyl chloride) (PVC), 109–111
Free radical polymerization
mechanism of, in ionic liquids, 10
polymer tacticity, 8–9

G

Gelation
ionic liquids and polymer network, 157, 158*f*
ionic liquids with single-walled carbon nanotubes, 166–167
solvents, 150
See also Poly(ethylene glycol) (PEG) networks; Single-walled carbon nanotubes (SWNTs)
Gel permeation chromatography (GPC), poly(n-butyl methacrylate) using imidazolium salts as solvents, 24*f*, 26*f*, 27*f*
Glass transition temperature
differential scanning calorimetry (DSC) traces of poly(methyl methacrylate)s (PMMAs), 9
effect of addition to solid polymer electrolytes, 98, 99*f*

effect of LiTFSI addition to ionic
liquid (IL) polymer, 93–94
imidazolium salts, 20, 21*t*
plasticized poly(vinyl chloride)
(PVC), 110*t*
poly(butyl methacrylate), and
degree of polymerization, 27–28
Green solvents, potential of ionic
liquids, 39

H

1-Hexyl-3-methylimidazolium
bis(trifluoromethanesulfonyl)imide
differential scanning calorimetry
(DSC) traces of neat or gelled, 153,
154*f*
functionalized poly(ethylene
glycol) oligomers, 72, 73*f*
See also Poly(ethylene glycol)
(PEG) networks
2-Hydroxyethyl methacrylate
(HEMA)
compatibility of 1-ethyl-3-
methylimidazolium
tetrafluoroborate and 1-
butylpyridinium
tetrafluoroborate with, and its
polymers, 125*t*
storage modulus and loss of tangent
for HEMA network with
dissolved ionic liquid vs.
temperature, 127*f*
temperature dependence of ionic
conductivity for HEMA network
polymers, 128*f*, 129*f*
See also Polymer electrolytes

I

Imidazolium salts
impurities, thermal properties,
recrystallization temperature,

melting temperature, and
viscosity, 21*t*
polymerization of 3-(*N*-2-
(methacryloyloxy)ethyl]-*N*,*N*-
dimethyl-3-ammonio)
propanesulfonate, 19–20, 29–32
polymerization of butyl
methacrylate using, 24–28
polymerization of monomers, 19–
20
recycling of ionic liquids after
polymerization, 32, 33*f*
relation between molecular weight
and structure, 27
structure, 20, 21*t*
thermal degradation, 23*f*
thermal properties and viscosities,
22, 23*f*
Impurities, imidazolium salts, 19–20,
21*t*
Industrial polymerization
methyl methacrylate (MMA), 43,
46*f*
PMMA production using water-
soluble ionic liquids, 46, 47*f*
Interaction parameters, solvents, 76*t*
Ionic conductivity
effect of addition to solid polymer
electrolytes, 98, 99*f*
effect of LiTFSI addition to ionic
liquid (IL) polymer, 93–94
ion gels, 128–131
spacer for IL polymer, 92
temperature dependence for IL
polymer with varying LiTFSI,
92–93
Ionic liquids (ILs)
acid catalysis for polymerization,
72
advantages as plasticizers, 104,
105*t*
applications, 72, 90
cellulose solubility, 73–76
electrolytes salts for polymer-in-
salt systems, 120, 122

gelation of, with single-walled carbon nanotubes, 166–167
incorporation into gel matrix, 156–157
interaction parameters, 76*t*
polymer applications, 72
polymer electrolyte technologies, 72
potential as "green solvents", 39
potential replacement for volatile organic compounds, 38, 47–48, 150, 158
properties, 90
reaction media in polymerization techniques, 38
recycling after polymerization, 32, 33*f*
reverse atom transfer radical polymerization (ATRP) in, with ester group, 61, 62*t*
reverse ATRP of MMA in chiral, 61, 63–64
solution polymerization vs. bulk processes, 38–39
See also Cellulose; Crosslinked polymers in ionic liquids; Plasticizers for polymers; Polymer electrolytes; Reverse atom transfer radical polymerization (ATRP); Room temperature ionic liquids (RTILs); Single-walled carbon nanotubes (SWNTs); Solid polymer electrolytes; Water-soluble ionic liquids (ILs)
Ion transport behavior, ion gels, 128–131

L

Laccase enzyme
cellulose polymer films, 79–81
Leaching, plasticizers in plasticized poly(vinyl chloride), 114, 116–117

LiCl/dimethylacetamide, dissolution of cellulose, 74
Linear solvent-energy relationship (LSER), cellulose solubility, 75–76
Lithium salt, LiTFSI
effect on ionic conductivity and glass transition temperature, 93–94
effect on ionic conductivity of ionic liquid (IL) polymer, 92–93
fraction of "free" and "ion paired" TFSI anions, 95, 96*f*
Raman spectra of, and IL polymer, 94–95
structure, 91*f*
temperature dependence of ionic conductivity for IL polymer, 92–93
See also Solid polymer electrolytes
Living radical polymerization
reverse ATRP, 53
See also Reverse atom transfer radical polymerization (ATRP)
Loss moduli, bucky gels, 169, 171*f*

M

Magnetite/cellulose composites
preparation, 78
scanning electron microscopy (SEM), 79
X-ray powder diffraction patterns, 78
See also Cellulose; Composites
Maleimides, *N*-substituted, reverse atom transfer radical polymerization with styrene, 56
Mechanisms, free radical polymerization in ionic liquids, 10
Melting temperatures, imidazolium salts, 19–20, 21*t*
Methacrylates
transparency and glass transition temperatures, 18

See also Butyl methacrylate (BMA)
3-(*N*-2-(Methacryloyloxy)ethyl]-*N,N*-
dimethyl-3-ammonio)
propanesulfonate
experimental, 19–20
^{1}H NMR spectra of copolymers of,
with butyl methacrylate, 31*f*
^{1}H NMR spectra of polymer,
30*f*
polymers containing zwitterionic
structures, 29, 32
structure, 20*f*
See also Imidazolium salts
Methyl methacrylate (MMA)
characteristic data of PMMA and
polystyrene (PSt) in ionic
liquids, 11*t*
compatibility of 1-ethyl-3-
methylimidazolium
tetrafluoroborate and 1-
butylpyridinium
tetrafluoroborate with, and its
polymers, 125*t*
kinetic measurements of
polymerization in bulk and
water-soluble ionic liquids, 42–
43, 44*f*
polymerization in ionic liquids, 5,
6*t*
reverse atom transfer radical
polymerization (ATRP) in 1-
butyl-3-methylimidazolium
tetrafluoroborate and 1-dodecyl-
3-methylimidazolium
tetrafluoroborate, 59–60
reverse ATRP in 1-butyl-3-
methylimidazolium
hexafluorophosphate, 55–
59
reverse ATRP in chiral ionic
liquids, 61, 63–64
reverse ATRP in ionic liquids with
ester group, 61, 62*t*

thermal polymerization in ionic
liquid, 12
typical continuous industrial
process for polymerizing MMA,
43, 46*f*
See also Reverse atom transfer
radical polymerization (ATRP)
Migration, plasticizers in plasticized
poly(vinyl chloride) (PVC), 114,
115*t*
Miscibility, room temperature ionic
liquids with organic solvents, 2
Molecular weight data, polymerization
of methyl methacrylate (PMMA)
and styrene (St) in benzene and
ionic liquids, 6*t*, 7

N

Nanocomposites. *See* Polymer-carbon
nanotube nanocomposites;
Polymer-layered silicate
nanocomposites
Nanotubes. *See* Single-walled carbon
nanotubes (SWNTs)
Networks. *See* Poly(ethylene glycol)
(PEG) networks
N-substituted maleimides, reverse
atom transfer radical
polymerization with styrene, 56
Nuclear magnetic resonance (NMR)
^{13}C NMR spectra of PMMA in
benzene and ionic liquid, 8*f*
solid state NMR studies of
poly(divinylbenzene), 141–143

O

Organically modified layered silicates.
See Polymer-layered silicate
nanocomposites

P

Plasticizers
  di(2-ethyl hexyl) phthalate
    (DEHP), 104
  ionic liquid, 107*f*
  ionic liquids (ILs) as, 104
  mass loss during temperature ramp,
    112*f*
  objective of replacing traditional,
    105
  polymer formulations, 104
  structures and abbreviations for
    ionic liquid plasticizers, 107*f*
  structures and abbreviations for
    traditional, 106*f*
  thermal stability, 111–113
  traditional, 106*f*
Plasticizers for polymers
  characterization methods, 108–109
  chemical structures and
    abbreviations for traditional and
    ionic liquid (IL) plasticizers,
    106*f*, 107*f*
  experiment, 105, 108–109
  flexibility for plasticized poly(vinyl
    chloride) (PVC), 109–111
  flexibility, 109–111
  glass transition temperatures of
    PVC with traditional or IL
    plasticizers, 110*t*
  leaching, 114, 116–117
  long-term thermal stability of
    plasticized PVC, 113*f*
  mass loss of plasticized PVC
    samples by leaching, 116*t*
  migration of plasticizers in
    plasticized PVC, 114, 115*t*
  objective to replace traditional
    plasticizers, 105
  preparation of plasticized polymer
    samples, 108
  tensile moduli for PVC with
    traditional or IL plasticizers,
    111*t*

ultraviolet stability of plasticized
    PVC, 114, 115*t*
  uniformity of imidazolium IL-
    plasticized PMMA samples, 109
Polarity, ionic liquids and
  polymerization, 10, 11*t*, 12
Polyacrylonitrile (PAN)
  cellulose/PAN blends, 82–83
  solubility in ionic liquids, 81–82
  See also Acrylonitrile
Poly(ethylene glycol) (PEG)
  hydrogels, 150
  ionic liquid-PEG gels, 72, 73*f*
  solubility in ionic liquids, 81–82
Poly(ethylene glycol) (PEG) networks
  alcohol partitioning, 153, 159–
    160
  attempted gelation of various
    solvents, 157, 158*f*
  bulk properties of ionic liquid (IL),
    153
  conductivity of 1-hexyl-3-
    methylimidazolium
    bis(trifluoromethanesulfonyl)imi
    de ([C₆mim][NTf₂])-PEG, 158,
    159*f*
  DSC traces of neat and gelled
    [C₆mim][NTf₂], 153, 154*f*
  effects of salt type on shrink-swell
    response, 155–156
  experimental, 151–153
  free-standing gels, 157
  gel compositions, 152*t*
  IL incorporation into gel matrix,
    156–157
  ILs as organic phase replacements,
    158
  shrink-swell response behavior of
    IL-PEG gels, 155–156
  thermal stability of gels by
    thermogravimetric analysis,
    153–154
Poly(2-hydroxyethyl methacrylate)
  (PHEMA)
  solubility in ionic liquids, 81–82

See also 2-Hydroxyethyl
  methacrylate (HEMA)
Polymer-carbon nanotube
  nanocomposites
  dispersion and phase separation,
    185–186
  melting characteristics, 184,
    185*f*
  preparation, 184–185
  transmission electron microscopy
    (TEM), 186*f*
  X-ray diffraction, 184
Polymer electrolytes
  classification, 121*f*
  compatibility of 1-ethyl-3-
    methylimidazolium
    tetrafluoroborate (EMIBF₄) and
    1-butylpyridinium
    tetrafluoroborate (BPBF₄) with
    vinyl monomers and polymers,
    125*t*
  conductive nature of ionic liquids
    (ILs), 157
  EMIBF₄ and BPBF₄ as
    polymerization solvents, 124
  1-ethyl-3-methylimidazolium
    bis(trifluoromethane
    sulfonyl)imide (EMITFSI) for
    gel preparation, 122, 123–124
  in situ polymerization of vinyl
    monomers in ionic liquids, 122–
    123
  ion gels, 122–123
  ionic conductivity and ion transport
    behavior in ion gels, 128–131
  ionic liquids for polymer-in-salt
    electrolytes, 120, 122
  ionic liquids for preparation of ion
    gels, 123–124
  molecular structures of room
    temperature ionic liquids, 122
  photograph of, film of
    EMITFSI/PMMA ion gel,
    126*f*

radical polymerization of vinyl
    monomers in ionic liquids and
    characterization of, 124–127
  storage modulus and loss of tangent
    for 2-hydroxyethyl methacrylate
    (HEMA) network polymer with
    dissolved BPBF₄ vs.
    temperature, 127*f*
  temperature dependence of ionic
    conductivity for HEMA network
    polymers with dissolved BPBF₄,
    129*f*
  temperature dependence of ionic
    conductivity for HEMA network
    polymers with dissolved
    EMIBF₄, 128*f*
  temperature dependence of
    relaxation time in ion gels, 130*f*
  thermal properties and ionic
    conductivities for ILs, 124*t*
  unique properties, 120
  See also Solid polymer electrolytes
Polymerization media, room
    temperature ionic liquids, 2–3
Polymerization processes. solvents, 18
Polymer-layered silicate
    nanocomposites
  abbreviations for study chemicals,
    177*t*
  experimental, 176–177
  instrumentation, 178–179
  nanocomposite preparation, 178
  OMS (organically modified layered
    silicates) preparation, 177–178
  thermal stability, 179, 180*f*, 181
  thermal stability of polystyrene
    nanocomposites, 181, 182*f*
  thermogravimetric onset and peak
    temperatures and shifts in X-ray
    diffraction (XRD) peak for d-
    spacing, 182*t*
  transmission electron microscopy
    (TEM) of polystyrene
    nanocomposite, 183

XRD of organically modified clays, 181*f*

Polymer materials. *See* Cellulose

Polymers
polymerizations in ionic liquids, 134–135
solubility in ionic liquids, 81–82
See also Crosslinked polymers in ionic liquids

Polymer supported electrolytes
ionic liquid application, 72
See also Polymer electrolytes; Solid polymer electrolytes

Polymer tacticity, poly(methyl methacrylate)s in benzene and ionic liquid, 8–9

Poly(methyl methacrylate) (PMMA)
characteristic data of PMMA and polystyrene (PSt) in ionic liquids, 11*t*
^{13}C NMR spectra of, in benzene and ionic liquid, 8–9
differential scanning calorimetry of, in benzene and ionic liquid, 9*f*
polymerization in ionic liquids, 5, 6*t*
precipitation, 43, 46*f*
size exclusion chromatography traces, 7*f*
typical continuous industrial polymerization process, 46*f*
water-soluble ionic liquids allowing precipitation, 43, 46*f*
See also Reverse atom transfer radical polymerization (ATRP)

Polystyrene (PSt)
characteristic data of poly(methyl methacrylate) (PMMA) and PSt in ionic liquids, 11*t*
polymer-carbon nanotube composites, 184–186
polymerization in ionic liquids, 5, 6*t*

PSt-carbon nanotube nanocomposite preparation, 178
See also Styrene

Poly(vinyl chloride) (PVC). *See* Plasticizers for polymers

Pore-size distributions
poly(divinylbenzene)s in ionic liquid and toluene, 140*f*
poly(trimethylolpropane trimethacrylate) [poly(TRIM)] in ionic liquid and toluene, 144*f*

Porogens
effect on polymerization products, 144–145
ionic liquids as, 140–141
See also Crosslinked polymers in ionic liquids

R

Radical polymerization
advantages and limitations, 52, 54
background, 52–54
See also Polymer electrolytes; Reverse atom transfer radical polymerization (ATRP)

Raman spectroscopy
ionic liquid (IL) polymer as function of LiTFSI concentration, 94–95
zwitterionic liquid addition to solid polymer electrolyte, 97, 98*f*

Recrystallization temperatures, imidazolium salts, 19–20, 21*t*

Recycling, ionic liquids after polymerization, 32, 33*f*

Reverse atom transfer radical polymerization (ATRP)
advantage of, of methyl methacrylate (MMA) in ionic liquid, 58–59
chiral ionic liquids, 61, 63–64
experimental, 54–55

ionic liquids with ester group, 61, 62*t*

mechanism at ATRP, 52

mechanism of reverse ATRP, 53

MMA by, in 1-butyl-methylimidazolium hexafluorophosphate [bmim][PF$_6$] at different [AIBN]/[CuCl$_2$] ratio, 58–59

MMA by, in [bmim][PF$_6$] at different temperatures, 56, 57*t*

MMA in 1-butyl-methylimidazolium tetrafluoroborate [bmim][BF$_4$] and 1-dodecyl-3-methylimidazolium tetrafluoroborate [C$_{12}$mim][BF$_4$], 59–60

MMA in chiral ionic liquids, 63–64

structure of PMMA by, in [bmim][PF$_6$], 57–58

Rheological properties, bucky gels, 169, 171*f*

Room temperature ionic liquids (RTILs)

abbreviations, 5*t*

applications, 2

characterization date for polymerization of methyl methacrylate (MMA) and styrene (St), 6*t*

^{13}C NMR spectra of PMMA in benzene and trishexyl(tetradecyl)phosphonium dicyanamide ([H$_3$TDP]Dcy), 8*f*

conversions in polymerization of MMA, 7

description, 2

diffusion-controlled termination, 10

differential scanning calorimetry traces of PMMA in bulk and in [H$_3$TDP]Dcy, 9*f*

experimental for polymerizations, 3–4

free radical polymerization mechanism, 8

instrumentation, 4

miscibility, 2

polarity, 10, 11*t*, 12

polymerization media, 2–3

polymerization method, 4

polymerization of styrene or MMA, 5, 7

range of cations and anions, 5

recycling after polymerization, 32, 33*f*

size exclusion chromatography of PMMA, 7*f*

solubilities, 4

tacticities in RTILs vs. organic solvents, 8–9

thermal polymerization of MMA, 12

viscosity, 10, 11*t*

See also Imidazolium salts

S

Salt types, shrink-swell response of gels, 155–156

Scanning electron microscopy (SEM)

cellulose fiber, 75*f*

cellulose/polyacrylonitrile blends, 83

magnetite/cellulose composites, 79

poly(divinylbenzene)s in ionic liquid and toluene, 141*f*

poly(methyl methacrylate) with traditional and ionic liquid plasticizers, 109

poly(trimethylolpropane trimethacrylate) [poly(TRIM)] in ionic liquid and toluene, 144*f*

Shrink-swell response behavior

ionic liquid-poly(ethylene glycol) gels, 155–156

testing procedure, 152

Single-walled carbon nanotubes (SWNTs)
  angular frequency dependencies of dynamic storage and loss moduli of bucky gel, 169, 171*f*
  characterization of bucky gels, 167–168
  critical gelator concentration, 167*t*
  description, 164
  differential scanning calorimetry thermograms, 170*f*
  dynamic hardness measurement, 165
  electrical conductivity measurement, 165
  experimental, 164–165
  fabrication and properties of ionic liquid polymer/SWNTs composite, 172
  gelation of ionic liquids (ILs) with, 166–167
  ionic liquids (ILs) for processing, 164
  measurements, 165
  molecular formulas and structures of ionic liquids (ILs), 167*f*
  phase-transition behavior of bucky gels, 168–169
  rheological properties of bucky gels, 169, 171*f*
  transmission electron microscopy image, 168*f*
  X-ray diffraction profiles, 170*f*
Size exclusion chromatography (SEC), poly(methyl methacrylate) polymerized in benzene and ionic liquid, 7*f*
Solid polymer electrolytes
  calculation of fraction of "free" and "ion paired" TFSI anions, 95
  dependence on concentration of fraction of "free" and "ion paired" TFSI anions, 96*f*
  effect of adding LiTFSI on ionic conductivity of IL polymer, 92–93
  effect of addition of zwitterionic liquid (ZIL), 96–98
  effect of LiTFSI on ionic conductivity and glass transition temperature, 93–94
  effect of ZIL concentration on ionic conductivity and glass transition temperature, 98, 99*f*
  experimental, 90–92
  flexible spacer for maintaining conductivity during polymerization, 92
  methods, 91–92
  photograph, 97*f*
  preparation of novel polymer gel electrolytes, 91
  Raman spectra of LiTFSI and IL polymer as function of LiTFSI concentration, 94–95
  structure of IL monomer, crosslinkers, LiTFSI, and ZIL, 91*f*
  synthesis of IL monomer and other components, 90, 91*f*
  temperature dependence of ionic conductivity, 93*f*
  See also Polymer electrolytes
Solubilities
  cellulose in ionic liquids, 73–76
  room temperature ionic liquids, 4
Solvents, interaction parameters, 76*t*
Stereocontrol, tacticity of free radical polymers, 8–9
Stereostructure of polymer, radical polymerization, 54
Styrene
  compatibility of 1-ethyl-3-methylimidazolium tetrafluoroborate and 1-butylpyridinium tetrafluoroborate with, and its polymers, 125*t*

kinetic measurements of polymerization in bulk and water-soluble ionic liquids, 42–43, 45*f*
polymerization in ionic liquids, 5, 6*t*
precipitation on reaction in ionic liquid, 7
reverse atom transfer radical polymerization with *N*-substituted maleimides, 56
See also Polystyrene (PSt)

T

Tacticity, poly(methyl methacrylate)s in benzene and ionic liquid, 8–9
Tensile modulus, plasticized poly(vinyl chloride) (PVC), 110, 111*t*
Termination, diffusion-controlled, 10
Thermal polymerization, methyl methacrylate (MMA) in ionic liquid, 12
Thermal properties, imidazolium salts, 19–20, 21*t*, 22, 23*f*
Thermal stability
ionic liquid-poly(ethylene glycol) gels, 153–154
plasticized poly(vinyl chloride), 113
polymer-layered silicate nanocomposites, 179, 180*f*, 181
traditional and ionic liquid plasticizers, 111–113
Transmission electron microscopy (TEM)
polymer-carbon nanotube nanocomposites, 186
polymer-layered silicate nanocomposites, 183
single-walled carbon nanotubes in ionic liquid, 168*f*

Trimethylolpropane trimethacrylate (TRIM)
polymerization procedure, 137
See also Crosslinked polymers in ionic liquids
Trishexyl(tetradecyl)phosphonium dicyanamide ([H₃TDP]Dcy), ^{13}C NMR spectra of PMMA in benzene and, 8*f*
Trommsdorf effect, polymerization in ionic liquids, 10

U

Ultraviolet stability, plasticized poly(vinyl chloride) (PVC), 114, 115*t*
Uniformity, ionic liquid-plasticized PMMA, 109

V

Vinyl acetate, compatibility of 1-ethyl-3-methylimidazolium tetrafluoroborate and 1-butylpyridinium tetrafluoroborate with, and its polymers, 125*t*
Vinyl monomers, radical polymerization in ionic liquids, 124–127
4-Vinylpyridine (VP)
copolymerization procedure with divinylbenzene, 137
See also Crosslinked polymers in ionic liquids
Viscosity
dependence of polymerization degree on ionic liquid, 25–26
imidazolium salts, 19, 21*t*
ionic liquids and polymerization, 10, 11*t*

Volatile organic compounds (VOCs), potential replacement by ionic liquids, 38, 47–48, 150, 158

W

Water content, imidazolium salts, 20, 21*t*
Water-soluble ionic liquids (ILs)
  advantages as reaction media, 46
  butyl methacrylate (BMA) polymerizations, 42–43, 44*f*
  experimental, 42
  kinetic results of polymerizations, 42–43, 44*f*, 45*f*
  methyl methacrylate (MMA) polymerizations, 42–43, 44*f*
  objective of polymerization study, 39, 42
  physical properties, 40*t*, 41*t*
  polymerization processes of interest, 47
  potential replace common volatile organic compounds (VOCs), 39
  precipitation of PMMA, 43, 46*f*
  proposal for industrial production of PMMA using, 47*f*
  scheme of typical continuous process for MMA polymerization, 46*f*
  styrene polymerizations, 42–43, 45*f*

X

X-ray powder diffraction
  magnetite/cellulose composites, 78
  polymer-carbon nanotube nanocomposites, 184
  polymer-layered silicate nanocomposites, 179, 181*f*
  single-walled carbon nanotubes in ionic liquid, 170*f*

Z

Zwitterionic liquids (ZIL)
  effect of addition to solid polymer electrolytes, 96–98, 99*f*
  effect on ionic conductivity and glass transition temperature, 98, 99*f*
  photograph of solid polymer electrolytes by ZIL concentration, 97*f*
  preparation of novel polymer gel electrolytes, 91
  Raman spectra of ionic liquid (IL) polymer with ZIL addition, 97, 98*f*
  structure, 91*f*
  See also Solid polymer electrolytes
Zwitterionic structures, polymers containing, 29, 32

Printed in the United States
80763LV00002BA/26